Islamic Tourism

Islamic Tourism

Editor

Heydar Hussain

scitus
academics

Islamic Tourism
Edited by **Heydar Hussain**

ISBN: 978-1-68117-231-6
Library of Congress Control Number: 2016934763

www.scitusacademics.com
Box No. 4766, 616 Corporate Way,
Suite 2, Valley Cottage,
NY 10989

Preface

Tourism is a complicated concept that covers a wide range of social, behavioural, economic, political, cultural, and environmental considerations. The concept of tourism consists of a set of activities, services, and benefits which all give tourists particular experiences. Islam is considered a major tourism market globally; however, special values of Muslims – who have particular needs in their system of beliefs – have not been well defined in the tourism market. Islamic tourism is all marketing and product development effort that are directed at Muslims, even without completely religious motivations. Efforts that emphasize the significance of Muslim and non-Muslim tourists as new markets and destinations for tourism; In other words, such efforts try to promote tourism among Muslims, develop new tourism destinations, and strengthen inter-organization and inter-government cooperation in the Islamic World.

The religious-conservative concept for Islamic tourism has not yet been theoretically articulated. But various opinions and remarks in the discussions on the future of tourism in the Arab and Islam worlds as well as some practices of hotel's managements indicate that articulations and implementations are just a matter of time. The whole idea is based on the conservative interpretation and understanding of Islam. Merging elements of the extremely conservative Islamic lifestyle with the modern tourism industry could indeed present new tourism options, spaces, and spheres. Excluding the fundamental isolationistic views which reject practically any tourism activities beyond pilgrimage, the religious-conservative concepts may be described as "community tourism". Islamic resorts and hotels as well as Islamic destinations and programs are shaping some tourist activities; Islamic resorts funded and owned by Islamic finance institutions, alcohol-free accommodations, no in-door disco/nightlife hotels, gender-segregated fitness and sport facilities, conservative in-door dress-codex, availability of prayer-rooms on site, women floors and "Islamic" entertainment programs. Many elements of this concept are indeed implemented in the domestic tourism inSaudi ArabiaandIran.Halal tourism is a subcategory of tourism which is geared towards Muslim families who abide by rules of Islam. The hotels in such destinations do not serve alcohol and have separate swimming pools and spa facilities for men and women. Malaysia, Turkey and many more countries are trying to attract Muslim tourists from all over the world offering facilities in accordance with the religious beliefs of Muslim tourists. The Halal tourism industry also provides flights where no alcohol or pork products are served, prayer timings are announced, and religious programs are broadcast as part of entertainment offered on board. The cultural concept for Islamic tourism includes visions and ideas that outline the inclusion of Islamic religious-cultural sites in tourism programs with "pedagogical" and self-confidence-building elements. It tries to encourage a re-orientation inside the tourist destinations towards less consumption and "western-culture" loaded sites towards more Islamic historical, religious and cultural sites. The economic concept for Islamic tourism is an extension and expansion oriented concept which focuses on the importance of intra-Muslim and intra-Arab tourism in terms of inclusion of new tourist markets and tourist destinations. The economic concept is the most wide-spread and widely discussed in the Arab and Muslim worlds at different levels. It considers the Muslim countries as one of the emerging tourist markets of the future with huge economic, demographic and destination potential.This book, Islamic Tourism, focuses on such issues as engagement (by Muslims), places (Islamic destinations), products (residential places, foods, and beverage), dimensions (economic, cultural, religious, etc), and managing service processes (marketing and ethical issues).

Table of Contents

Chapter 1

Towards Sustainable Tourism Development in Zambia: Advancing Tourism Planning and Natural Resource Management in Livingstone (Mosi-oa-Tunya) Area

Binyi Liu, Floyd M. Mwanza

Department of Landscape Studies, Tongji University, Shanghai, China.

ABSTRACT

Over the last few decades, development policy has been dominated by mainstream economic theories that focus on economic growth to achieve sustainable development. The pace and scale of tourism growth in Livingstone (Mosi-oa-Tunya) area in Zambia have seen over reliance on natural resource utilisation by mass tourism developments. Compounded by insufficient planning and limited co-ordination and collaboration among the institutions involved in the tourism sector, tourism can have a negative impact and can create conflicts. Tourism growth in Livingstone (Mosi-oa-Tunya) has predominantly focused on the economic incentives in tourism and ignored the social perspective and impact on the local population. In general, the government agency administration structures affect the successful implementation of tourism policy and planning for sustainable tourism development. Given the fact that the limited government support, funds and appropriate knowledge in tourism limit Livingstone (Mosi-oa-Tunya) to develop as a sustainable "green" destination and remain an enormously difficult task to achieve.

Keywords: Natural Resources; Planning; Sustainable; Tourism Policy

1. INTRODUCTION

In an effort to reduce the negative impacts of conventional tourism, more environmentally and socially conscientious approaches were promoted to tourism. Typically called "ecotourism and sustainable tourism" though other terms such as responsible tourism, nature based tourism, green tourism, and alternative tourism are also used [1]. Any tourism destination without an adequate plan for development that addresses the economic as well as social and environmental functions of the industry is under prepared for the impacts of visitors, catastrophic events, and enforcing market forces [2]. Tourism requires a great deal of infrastructure including hotel road parking lots and restaurants which typically brings a number of negative consequences [1], such as increased pollution levels, the destruction of natural habitats, the displacement of natural wildlife and undesirable influences to once remote cultures [3]. As an alternative to conventional tourism, sustainability and ecotourism has continued to gain momentum over the last two decades [4]. Planning for sustainable tourism development refers to environmental preservation planning and as such includes a variety of inquiry activities and analysis to the decision for determining the direction of the development [1,5]. Tourism planning is advanced to prevent the intensive utilisation of resources in some specific areas without previous care for the preservation of the resources [6]. There has been an increasing need for landscape planners to consider methodological approaches to tourism planning and a number of techniques, principles, and examples that have evolved and recommended [7-9]. Nevertheless, the multiplicity and heterogeneity of tourism stakeholders render the process complicated [10]. A key component to the success of sustainable "green" tourism is local control in the planning, development and management of these tourism sites [11]. Arguably, Livingstone has the highest concentration of tourism activities in Zambia [12], although many authors have examined various aspects of general economic planning viability. A literature review of tourism sector shows that, with few exceptions, most studies in the past have focused on research in tourism marketing [13], and that many now incorporate new theories and concepts for better provision of tourism [14]. Bearing in mind the above, this study aims to contribute to better understanding of how tourism development planning is carried out and the inherent difficulties are associated with the implementation. It aims to explore and examine what is happening in Zambia. Therefore, this study attempts to record and analyze the factors affecting the development of the Livingstone (Mosi-oaTunya) tourism region. It compares planning and implementation of the national and regional tourism policies and strategies and attempts to show the conditions, which affect the course of implementation and cause its divergence from sustainable national policies and regional planning objectives. It considers the processes of translating objectives to outcomes and investigates why those processes fail to translate many objectives to practice.

2. LITERATURE REVIEW/BACKGROUND INFORMATION

First, Zambia's tourism industry relies on two primary assets: the Livingstone (Mosi-oa-Tunya) area and the country's wildlife estate in 19 national parks, Game Management Areas (GMAs) and game ranches Zambia Development Agency [15]. The Livingstone (Mosi-oaTunya) Falls site appeals to a large tourists range of approximately 138,830 visitors more than the safari product of 61,000 visitors in 2009 Ministry of Tourism Environment & Natural Resources & Zambia Tourism Board [16]. Livingstone tourism activities are relatively far developed, compared to other regions in Zambia, [15]. Livingstone Mosi-oa-Tunya has a large proportion of Zambia's adventure tourism capacity [17]. A recent [18] publication elaborates that between 2010 and 2030 arrivals to emerging economies will increase at double the rate. As a result, the market share of emerging economies such as Zambia has increased from 30% in 1980 to 47% in 2011 and expected to reach 57% by 2030 to over one billion international tourist arrivals [18]. However, to meet sustainable tourism, scholars argue that sustainability has largely been used conceptually as a "good idea" but has been difficult to enable through specific initiatives [4]. The task is more difficult in view of the multiple crises faced by the world. Recession, climate change, fuel crisis, food crisis, and water crisis, planning and governance become topical issues [19]. Consequently, the impetus for many of the current initiatives in tourism and international development stems from Agenda 21, a comprehensive program of action for attaining "sustainable development" in the 21st century [20].

International tourism destinations particularly those rich in biodiversity have in recent years caught attention of the global environmental movement because of resource degradation [21].

Arguably, community based planning approaches are promoted for tourism development as a prerequisite to sustainability [5,20]. Reference [22] observed that many destinations are now pursuing strategies that aim to ensure a sensitive approach when dealing with tourism. [8] refers to sustainable tourism development is an enormously difficult task to achieve in developing countries, without the collaboration of the international agencies such as the World Bank and the International Monetary Fund. Reference [23] observed the rise of sustainable tourism discourse in Southern Africa and has seen the development of a multiplicity of tourism projects packaged under ecotourism as a more sustainable form of tourism than mass tourism. The term "sustainable tourism" can mean different things to different people, often according to the position of the individual stakeholder. It is important to elaborate tourism planning with a definition of some principles of sustainable tourism [24,25]. More recently, the WTO defined "sustainable tourism" as follows: "Sustainable tourism development meets the wishes of present tourists and host regions while protecting and enhancing opportunities for the future." As [7] stated, achieving sustainable tourism is a continuous process and it requires steady monitoring of impacts, introducing the necessary preventive and remedial measures whenever necessary [25].

Furthermore, sustainable tourism is now an approach at the international level that is advocated to be adapted to prepare all types of tourism to be environmentally, socially and economically beneficial [26-28].

Tourism in Zambia

Zambia's protected area network covers 30% of the country (224,075 km^2) for which Zambia Wildlife Authority (ZAWA) is responsible [12].

The 19 National Parks covering 6587 km^2 (28%) and 32 Game Management Areas (GMAs) 160,488 km^2 in extent or 72% of the country's PA network a huge resource forms wilderness tourism supply side [29-31]. Many of the tourism activity centres on the 19 National Parks covering 63,587 km^2 (28%) and 32 Game Management Areas (GMAs) of 160,488 km^2 in extent, or 72% of the country's protected area (PA) network [29]. Many national parks landscapes and fauna form the basis for lucrative tourism and hunting industries in Zambia [30].

According to the [32] tourism activity created 44% of employment in the hotel and restaurant industry compared to 7% in the mining industry, 99% in agriculture and 66% in manufacturing. In 2010, the number of arrivals in Zambia was 815,000 and increased to 920,299 in 2011 as shown in Figure 1.

Zambia's tourism industry established itself in the 1950s, [17]. As shown in Table 1 Zambia's tourism indicators in years. There have been some significant changes in strategic and policy levels in Zambia, all of which have the potential to influence the sustainable tourism planning agenda [33].

However, the extent to which these changes have infiltrated into implementation of local government is an area that requires further investigation [34,35] and the holistic involvement of communities in effective utilisation of their environmental assets and cultural heritage. The basis for this stance stems from the factor that tourism is increasingly Livingstone major economic activity. According [16] Zambia's stake in the industry has been insignificant, but the past five years or so have witnessed a steady growth in the tourism sector, projected to deliver over 1.4 million tourist arrivals by 2015. [17]. Following Figure 2 provides more detail on the purpose of holiday visits, suggesting that more than half Zambia's holiday makers (54%) arrive with the intention of visiting Livingstone's Mosi-oa-Tunya only [36].

Zambia's major tourism supply side clusters have developed in only a few key urban and national park locations, with a strong bias to the Livingstone region that offers nearly 40% of all nature tourism [15]. The effects of uncontrolled tourism development degrade ecosystems can be negative [37,14]. Nowhere in Zambia is this more evident than in Livingstone Victoria Falls (Mosi-oatunya) tourism site [38]. This underscores the need to entrench sustainable tourism planning principles in tourism management plans well before development begins and irreparable damage are incurred [39,40].

3. RESEARCH SITE PROFILE AND CHARACTERISTICS

3.1. Study Site Location

Livingstone (Mosi-oa-Tunya) was purposefully selected

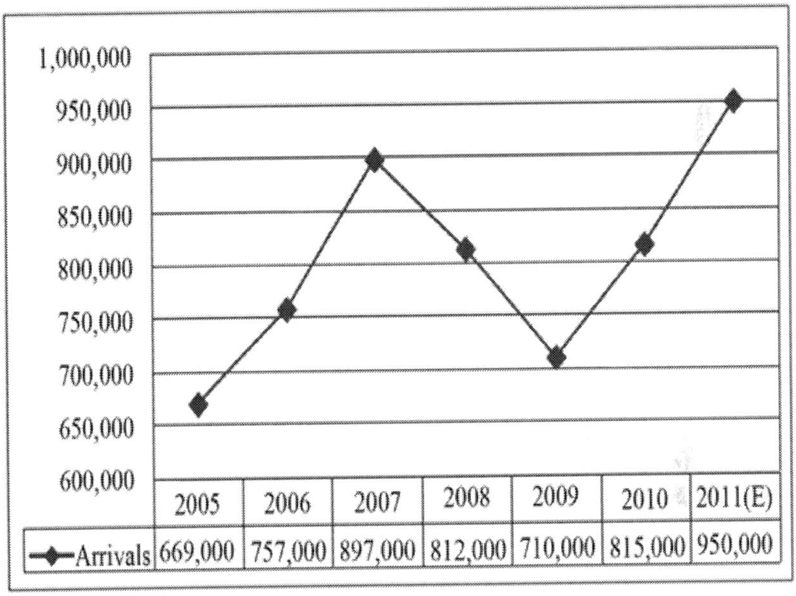

Figure 1. Tourists arrivals in Zambia as of source; [16].

Table 1. Zambia's Tourism indicators in years. Source: [34].

Year	Budget Allocation in (USD)	Budget Release in (USD)	Employment Levels	GDP (%)	Tourism Earnings (USD)	Tourist Arrivals
2006	8,140,000	3,202,000	21,204	2.40%	177,000,000	757,000
2007	34,740,000	21,740,000	22,204	2.00%	188,000,000	897,000
2008	24,264,000	13,616,000	22,756	2.40%	200,000,000	812,000
2009	15,520,000		24,308	2.70%	212,000,000	710,000
2010	43,990,000	13,260,000	25,960	2.30%	224,000,000	815,000
2011	7,120,000	6,320,000	31,900		216,000,000	950,000
2012	10,520,000					
2013	12,760,000					

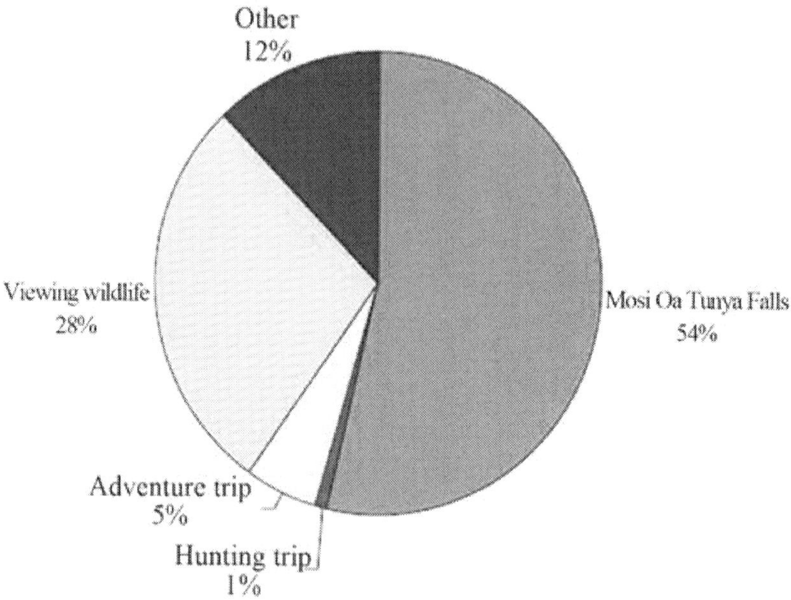

Figure 2. Tourists' destination choices in Zambia [17].

for the study because; it is among the sites being developed under the Livingstone Tourism Plan in Zambia [41]. Furthermore, the area is the earliest to be established in the region as a sustainable destination [42] and as a result, was due for evaluation.

The region derives its name from the waterfall locally known as the "Mosi-Oa-Tunya" meaning "The Smoke that Thunders". David Livingstone from which the Livingstone town is named after then named the waterfall "Victoria" after Victoria, the British Queen [29]. The size of the park is about 6555 hectares or 66 square kilometers lying between latitude 17°49' South to 17°54' South and longitude 25°41' East to 25°55' East and an attitudinal range of between 900m to 925m above sea level [41].

3.2. Demography and Socio-Economic Characteristics

Livingstone Victoria Falls (Mosi-oa-Tunya) is in Southern Province of Zambia. Established in 1971 and declared a National Park in 1972 [42]. With an estimated population of 136,897 [32] and approximately has over 6000 visitors per day [16]. They contain biodiversity of global significance and is listed as Critical Sites (including Critical Habitats) [43]. There are around 4590 plants confined to this area, together with 35 endemic mammals, 51 endemic birds,

52 endemic reptiles, 25 endemic amphibians and an unknown number of endemic invertebrates [43,44]. It is approximately 20 km long and a maximum of 5 km wide [41]. It is constricted centrally to approximately 0.5 km (500m of land) [41]. Figure 3 shows the location map for Livingstone Mosi-oa-Tunya area; One of the overriding concerns about tourism in Zambia is that the tourism product relies heavily on the natural and physical environment [42,45,46]. Reference [8] highlighted that unsustainable tourism activities can affect the future viability the tourism sector, conserving of natural resources has become important through planning [47,48]. "Governments have become extremely canny in reproducing the sustainable development rhetoric without actually effecting fundamental policy and legislation changes [20]." In Zambia the need to understand the impacts of tourism has become important within a planning context because of the many statutory requirements such as the [41,49,50] and global demand for sustainable tourism [25].

The Zambian tourism sector is guided directly and indirectly by 11 pieces of legislation [51]. These tourism plans have focused merely on maximising foreign tourists' receipts and thus increasing the supply capacity of the tourism industry [14,52]. The main shortcomings are due to sectorial planning done in isolation, communication and co-operation among related bodies are sparingly weak and in most cases do not exist [34,53-56]. It is only right that development and land management is supported by a holistic planning [57-60]. It is reported that these common shortcomings in present tourism development approach pose challenges to sustainable tourism development in Zambia [35,54,61,62]. As observed by [8,63-65] the study is premised on the assumption that, local government agencies and communities to influence the sustainable tourism planning agenda. However, the extent to which tourism policies has infiltrated into local tourism agencies and communities is an area that requires further investigation. The following section outlines the methodology used to survey local agencies and communities to ascertain responses to tourism policy and planning implementation in Livingstone (Mosi-oa-tunya) area.

4. METHODS AND PROCEDURE

4.1. Research Design

The research conceptual framework was developed based research methods for assessing local authorities participation in tourism policy and planning similar to studies done by [10,63,64]. Given that the purpose of this paper is to identify and evaluate tourism policy and planning implementation for sustainable tourism development by government tourism agencies in Livingstone (Mosi-oatunya) area. A qualitative descriptive approach was employed and quantitative data where appropriate. The study used the non-probability sampling design to collect data from local tourism authorities and agencies. As cited in [66], the purposive sampling technique was found to be adequate and appropriate for such a survey research. In view of the facts given

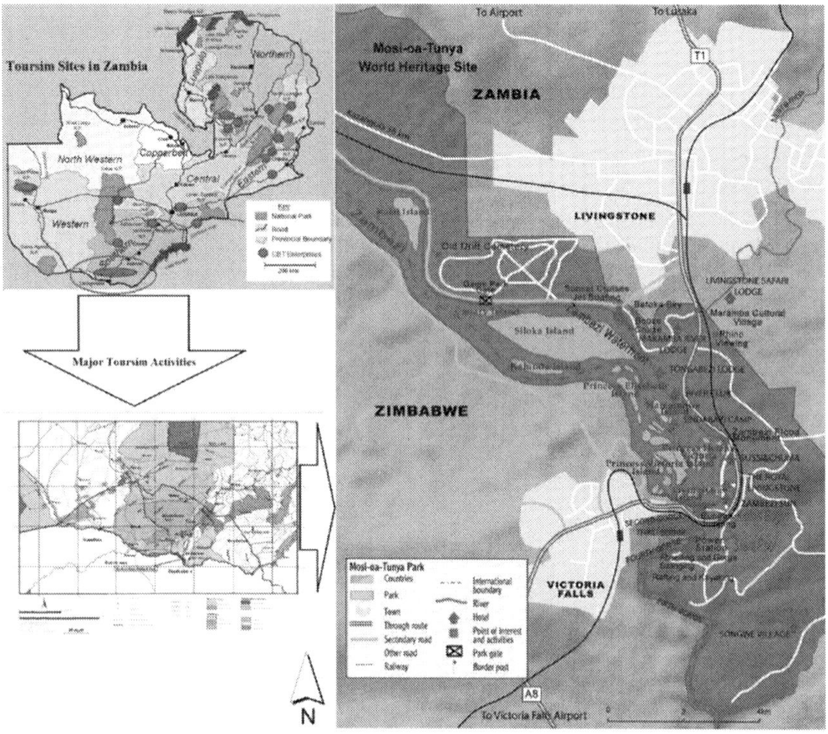

Figure 3. Map of Livingstone [35] (Mosi-oa-Tunya) of Zambia, source: modified by author (2013).

above, the purposive sampling method was adopted. Interview guides and questionnaires were the instruments used for data collection. The interviews and questionnaire administration was made to government tourism agencies (MTENR/ZAWA/ZEMA) and local Community-Based Natural Resource Management (CBNRM) representatives. In addition, institutions related to the Zambian tourism industry were also contacted for requisite information and data. To capture a significant number of tourism planners in the population sample, data were collected from tourism agencies at the Lusaka Ministry of Tourism Environment Natural and Resources, Zambia Wildlife Authority in Lusaka's Chilanga headquarters, Zambia Tourism Board Lusaka and Regional Local ZAWA branch in Livingstone (Mosi-oaTunya) area. Livingstone greater area community leaders of Community Resources Boards (CBR) agencies and popular lodges, tourism enterprises and guesthouses make the local community group. The research was conducted from 22nd November 2012 to 5th April 2013.

All the in-depth interviews were conducted at places of choice by the interviewees in the various departments and communities. The interviews were conducted by the corresponding author.

4.2. Data and Sources

In view of the facts given above, the purposive sampling technique was found to be adequate and appropriate because there was no sample frame of all respondents. Since it was an exploratory study, the rationale of the data collection was to ascertain government agencies and community heads' roles in tourism policy and planning participation in the sustainability framework implementation for the case of Zambia's Livingstone greater area. The survey employed three major methods, personal questionnaires, interviews of identified key actors in the Zambia tourism planning process and the local tourism businesses and local tourism authorities. The purpose however, was to generalize from a sample to the population in order that inferences could be made about the involvement of government agencies and th communities in tourism policy and planning development [67]. Secondary sources from literature review of books, journals and grey reports. Grey reports were used due to limited research publications concerned with tourism planning in developing countries such as Zambia. Past reports and any other material related to plans and policies for the tourism development of Livingstone were reviewed.

4.3. Sampling Procedure and Study Instruments

Based on the actors' group, a list of government tourism heads and local community representatives was compiled and used as a sampling frame for the selection of the respondents. A self-completion questionnaire survey was emailed to solicited sample list of local tourism authorities and planners in Zambia, who have direct and indirect involved in the Livingstone tourism master plan. A realtime online-based questionnaire website [46] was used to improve response and analysis of findings. The onlinebased survey enabled a degree of tracking and gauge findings and easy clarifications and adjustments in cases where the questionnaire was not well-defined. To encourage survey completion and to confine the aims of the survey to specific tourism planning objectives (such as identification of sustainable tourism planning implementation and development in tourism strategies), without eroding the aims of the investigation, the survey design incorporated a combination of closed and open questions which was also hosted online to improve survey responses and participation. Closed questions were utilised to gauge responses to straightforward questions, where a simple tick box suffices to assist in classification of respondents. However, recognizing the small sample specific population involved in this survey, a range of open questions were included to generate a source of more qualitative, explanatory information that can add a richer dimension to understanding responses. Hence, the fieldwork aimed to interview representatives of the major groups. It was designed using a series of semi-structured interviews with key actors. Reference [68] explained the criterion used to determine sample size is an important issue in research. The study uses descriptive data analysis and explanation, and the use of appropriate theory to help explain events [69,70].

4.4. Assumptions and Limitations

The study was based on the assumption that surveying experts with knowledge of the case study sites would be a reasonable way to obtain an up to date overview of site. However, this approach may have a few limitations, including:

- Bias of the expert's personal opinion;
- Interpretation of other stakeholder's views by a third party;
- Incomplete knowledge of the site (including of impacts, challenges and dynamics between stakeholders);
- Some experts had not been back to the site in the last two years.

This assumption and limitation were in part addressed by the desktop review of documents, which often served as a complement to the information provided by the experts.

4.5. Target Population and Sample Size

The target population for the study was government tourism agencies and local community heads or their representatives in the selected communities. A total of 85 questionnaires from 165 sent questionnaires were filled in for this particular study, 9 in-depth face to face interviews were carried out with persons involved in the policy making process and the implementation of tourism related plans. As shown in Table 3 the 85 respondent population comprised 12 local communities (CBNR) and consultants and 73 tourism planners/local authorities as shown in Table2 Surveys were mailed directly to the planning officers or agencies (MTENR, ZAWA, ZTB, ZEMA and National Heritage Conservation Commission (NHCC) who oversee on tourism planning processes and understand how tourism fits into local development plans for completion. 43 respondents completed and usable questionnaires were returned, yielding a response rate of 51%. The aim of the survey was not to produce large amounts of statistical analysis, rather to generate a picture of current levels of sustainable tourism planning at the two levels (local and regional), which is descriptive in exploring the small population size. Given that this figure represents half of all local authorities, the information the survey yielded is considered valid in providing a general picture of sector responses to tourism planning in Zambia.

4.6. Attitudes towards Tourism Development

The patterns of response provide a useful geographic spread of data, and represent a good mix of respondents. The low response rate at community level at 25%, explained primarily by the apparently delegated role of tourism planning by the local level of regional agencies and cross cutting issues, and whose main concerns relating to tourism are integrating resource management issues under (ZAWA/ZEMA/MTENR), particularly water and waste management. The respondents and their abilities to conceive the questions and to answer precisely varied not only from one group to another, but sometimes from one respondent to another in the same group. The gap in the level of knowledge, experience, and tourism planning development backgrounds and quality of information influenced the answers and the views of the respondents and thus the results obtained. At regional or both provincial and city levels, the response rates were over one half of the population (55%). Methodologically, this study suffered from the same problem as most online and mailed surveys, and while the overall response rate is satisfactory (often online and mailed 30% response rate is deemed reasonable for such surveys, for an online/ mailed survey, conventionally, a response rate of 20% is considered as a good response rate, while a 30% response rate is considered to be good [71]. It is difficult to assure the representativeness of the responses achieved. The non-respondents included 7 (Consultants & Planners) (out of 15) and 26 (local community & district planners; while for (ZAWA main branch) and regional planning offices the non-responsive figure was 9 (out of 12). Some 28 responses were received from MTENR Lusaka 49% response), while 15 responses were received from the Livingstone (54% response). Overall, the responses received provide a satisfactory sample in relation to tourism areas, population size and geography, all of which will be further elaborated in the findings. Longitudinal comparisons are only possible at the general level, given that although the same population was sampled, not all respondents answered the surveys. Lastly, it should be noted that, the names of specific government tourism departments are not given in the discussion of findings from the survey to respect the confidentiality of the research process, which was assured in the research process in order to get accurate responses.

5. RESULTS AND DISCUSSION

The findings of the survey are reported using a combination of descriptive and quantitative data given the small population, with verbatim responses to open questions to enrich the data and provide further insights. As a first step, it is valuable to recognise the scale and type of communities, areas and tourism profiles represented in these findings, particularly as such variables are useful in cross tabulating findings. The resident population of the survey areas at present population was estimated at 136,897 inhabitants at the 2010 census [32]. With reference the surrounding areas that make the largest populations, are made up of Mukuni's village on the eastern and South-Eastern border, Sekute Chiefdom (Simonga area) on the West, Imusho village to the western boundary of the park and Chief Musokotwane on the North-Western boundary are in Southern Province of Zambia, with an average population density of below 15 people km^2 residents with a total estimated population of 778,740 persons in Southern province alone [32,30].

5.1. Tourism Policies

Local and regional authorities were asked if they had knowledge of the Zambia Tourism Policy. There is no statutory requirement for a Tourism Policy, the publication of one indicates a strong community interest and local government commitment to tourism, the survey revealed that 26 tourism institutions under at the three levels of planning and implementation level have knowledge of the tourism policies and other strategies. Table 3 shows the comparison of the survey groups indicating the different types of organisational levels and knowledge on tourism policy and related strategies developed towards the tourism sector in Zambia.

The survey respondent's percentage outcome based on proximity to the study site revealed a lower understanding of the Tourism Policy. The respondents' percentage figure trends shows 33% of CBNRM respondents had knowledge of the policy and the trend rise in the knowledge of Tourism Policy and strategies by a significant rise at the main government ministry of tourism and government department agency ZAWA. This would appear to indicate that the effect of the national tourism strategy better understood at the core ministry and department and less appreciated or limited knowledge at local community level to develop and adopt strategies. Respondents with no knowledge of the Tourism Policy stated that all tourism matters delegated to the MTENR or ZAWA. Findings suggest that despite major tourism activities taking place in Livingstone and its surrounding areas, the local population have never come across the earmarked Livingstone greater area plan for sustainable development, but could be encouraged if they had one. This could explain the reason for low response from at local community level, and indicates a lack of interest in tourism development issues at this level, where tourism planning and policy issues were delegated to other bodies at ZAWA and MTENR.

5.2. The Influence of the MTENR/ZAWA Tourism Policy for Zambia (TPZ)

The majority of local tourism authority planning officers at local and government department (MTENR/ZAWA) level who had knowledge of the already existing tourism plans and had seen the Livingstone greater area plan, 83% of respondents had indicated how the Tourism Policy for Zambia (TPZ) would inform their own policy development. Responses from the local tourism authorities thought that there were emerging tourism issues that needed to be included in a future revised policy such as ecotourism certification, "green tourism", and sustainable tourism practices. Five respondents indicated the need to incorporate elements of the five year national development plans such as the [53,72,73], where appropriate to reflect particular locality and for easy implementation. While a further three stated that, they would take the plan lacked in implementation process due to institutional limitation and resources considered.

Table 2 . Composition of sample population responses of key tourism actors in Zambia.

	Year of Survey, 2012		Categories of Respondents			
No	Type of Organisation (s):	Target Pop:165	Respondents	Response	Non-Response	% Response
1	Line Ministries (Ministry of Tourism, Environment and Natural Resources (MTENR)			40	33	55
2	Zambia National Tourist Board (ZTB)	Government Officials				
3	Zambia Wildlife Authority (ZAWA)					
4	National Heritage Conservation Commission (NHCC)		73			
5	Zambian Environmental Management Agency (ZEMA)					
6	Donor Agency	Consultants				
7	Businesses &Parastals					
8	Tourism Experts					
9	Community GMA Game Management Area Community-Based Natural Resource Management (CBNRM)	Interest and Local Groups	12	3	9	25
	Total		85	43	42	80

Table 3. Tourism government agencies & actors' respondent's percentage response, by author (2013).

Year of Survey	Percentage tourism policy @ Community (CBNRM)	Percentage Regional offices tourism policy@ (NHCC/ZAWA)	Percentage tourism policy @ (ZAWA/MTENR)
2012	33%	66%	75%

Two respondents stated that the TPZ was approved in 1997 and published in 1999, does not reflect the current organisational structures and governmental existing plans, while a further three stated that the turnaround strategy plan directly aligns with the national sustainable strategies [12]. Others commented on specific elements of the national plan and appreciated the opportunity to determine the national context and direction of tourism strategy in Livingstone Zambia and the replicability of the pilot plans such as the Livingstone greater area development plan and for a common approach to core issues as set WTO's universal tourism standards. Overall, though, the ways in which the TPZ has already influenced, or will influence, policy at a local level appear vaguely stated in many cases.

5.3. Planning for Tourism Impacts in Livingstone

Some 57% of respondents raised specific tourism issues that need redress in the next review of the Tourism Policy Zambia. The responses as illustrated in Table 4 and in some cases, respondents gave more than one reaction. The range of emerging tourism related issues raised indicate two approaches to tourism development. These approaches are not polar opposites, but do represent different perspectives on tourism activity. On one side are those authorities that have concerns about the impacts of tourism, where key policy issues relate to balancing the needs of locals, visitors and other interests, dealing with impacts arising directly from tourism activity, and managing environmental resources (36% of authorities). A particular concern indicated by three council representatives is that of the cost of developing and managing tourism opportunities, activities and impacts. Two of these indicated impending studies to ascertain the economic cost of infrastructure and attractions, while a third noted the difficulties for councils with small populations to afford infrastructure improvements through the local rates system. 16% of respondents were conversely more concerned about developing tourism assets, promotions and infrastructure in an attempt to generate or meet the demand. Some 40% of respondent did not have any tourism issues of concern. The response may hide a number of more insidious issues, some local agencies do not possess the tourism expertise to identify and deal specifically with tourism impacts, while others may be more focused on championing the marketing orientation of Tourism in generating economic benefits, considering that policy literature focus on poverty reduction strategies. In many cases, there are significant dangers that negative impacts are not anticipated, mitigated or managed. Worth noting though is that 57% respondents identified tourism related issues that needed to be addressed the study. These findings indicate a growing interest and concern about the effects of tourism and the need for local tourism planning authorities to address impacts, both positive and negative, through the planning system. In addition, the range of issues identified, suggesting either a higher level of tourism awareness within councils or the emergence of a more extensive number of impacts.

Table 4 . Issues identified by respondents towards sustainable tourism, by author, (2013).

Issues	Number of Responses
Managing adverse environmental effects	3
Need to develop transport infrastructure	3
Waste disposal (especially relating to freedom camping)	3
Weighing up the economic cost of tourism	2
Conflicts between visitors, developers and residents	2
Product development	2
Concerns about effects of specific tourism developments	1
Different approaches adopted by different bodies	1
Addressing seasonality	1
Desire to maintain low impact tourism	1
Increasing demand for outdoor activities and how to meet it	1
Partnership and cultural opportunities	1
Oversupply of road stopping places	1
Effect of climate change on travel patterns	1
Increasing promotions	1
Pressure on infrastructure at peak times	1

5.4. Importance of Tourism in Livingstone

As indicated in [16,74] "74.9% of foreign tourists, who have had the opportunity to visit Zambia's popular tourism destinations", visited Livingstone Mosi-oa-Tunya area. Other popular destinations included South Luangwa National Park (24.3%), Mosi-oa-Tunya National Park (25.4%), Lake Kariba (26.5%) and other National Parks and Game reserves (26.5%) [16]. These destinations are Zambia's most developed and marketed attractions [75-78]. Respondents were asked to indicate if the perceived importance of tourism in the case of Livingstone. Some 50% respondents stated that the importance of tourism had increased, 17% of these stating increased significantly. The main reason given for this was the increasing recognition of the realized and potential economic benefits of tourism within the local tourism areas. It appears that many tourism stakeholders have become more aware of the beneficial effects that tourism can bring to a locality through as a source of revenue, business development and employment opportunities. In particular, the awareness of the ability of events to draw visitors to an area appears to have strengthened. Other contributing factors included growth in tourism, improved marketing and strategic vision, development of new products and services, and more central government funding. Only 7% respondents stated that the importance of tourism had decreased, partly due to the limited tourism appeal of one location but in two others a perceived lack of value, for example: Zambia Tourism Board ZTB have been unable to demonstrate, articulate and quantify the value in monetary terms. 26% of the respondents stated that the importance of tourism remained the same. This was explained by several locations where tourism activity remained static or where growth was limited by infrastructure constraints. One issue identified was the absence of effective tourism organisations and regional co-ordination to take tourism developments forward and to illustrate the benefits of tourism to the local communities, thereby not propelling tourism forward as a beneficial economic activity. Development of new attractions and recognition of substantial increases in visitors were cited as the main reasons for the increase in importance. This appears to indicate that tourism area have a clear understanding of how tourism can benefit their locality, which may have resulted from the key messages in the national tourism plan and associated reports. However, similar issues with regard to lack of financial support given to tourism or lack of importance placed on tourism activities.

5.5. Future Tourism Development

The range and scope of developments (78.6%) indicate a significant rise in the tourism infrastructure across the country, from airport enhancements to visitor trails. The most developments, which had taken place in three local communities in Livingstone (Mosi-oa-Tunya) area followed by accommodation development (non-hotel) in 30 areas (71.4%) [15]. It has been reported that the total number of tourists to Zambia is expected to reach more than 1.4 million tourists in by 2015 and these will require more hotel establishments in the country [15].

Table 5. Tourism planning pressures created in Livingstone greater area, by author (2013).

Tourism Pressures	Number of Respondents
Accommodating more visitors	5
Demands on local services	3
Anti-social behaviour/community spirit	2
Demands on water	2
Effects on wildlife	2
Waste volumes	2
Costs of stopping inappropriate development	1
Ensuring development no environmental degradation	1
Housing affordability for local residents	1
Increased freedom camping	1
Lack of workforce in peak season	1
Need to build more accommodation	1
River/waterfront subdivision	1

The development of new attractions at all levels suggests vibrancy in tourism development. In terms of the types of new developments, the list of new attractions, facilities and services on offer is considerable and far too extensive to include, but incorporates a large proportion of new trails, tours, guided walks and outdoor adventure activities, with a smaller amount of development to create or upgrade cafes, hotels, museums and retailing. All of which utilise environmental resources and all of which have the potential to create or exacerbate adverse impacts. As such, the role of the ZAWA in controlling the effects of tourism development is clear in a climate where growth in individual adventure tourism enterprises and outdoor pursuits is occurring. Some 44.2% of respondents considered local communities in the Livingstone area to be under pressure from increased tourism and Table 5 show the major pressures highlighted by respondents. Three broad categories of responses are distinguishable through examination of a subsequent open question on what pressures existed in localities.

First, specific locations were identified as likely to experience increased visitor numbers and associated impacts, e.g Mukuni's village on the eastern and Inyambo local tourism Community Development Trust areas. Second, the concerns arising from increased visitor numbers were identified including, demand for infrastructure, construction of tourist-related ventures, dealing with municipal waste, water demand and waste water disposal.

Increased freedom camping and effects on wildlife and natural areas, housing affordability, second homes and subsequent loss of community culture attributes, increase in tourist arrivals (e.g. Livingstone's newly extended Harry Mwaanga Nkumbula International airport expansion).

Third, and somewhat in contrast to the latter responses, a grouping of respondents though smaller than the latter, want to grow tourism and maximise the benefits, through creating infrastructure, building more accommodation and increasing the workforce. The survey identified that respondents in areas with the largest number of guest at night of over 5500 in the peak month were more likely to report that their area was under pressure from tourism. Correspondingly those with the smallest number of nights (less than 1000) were the least likely to be under pressure..The areas under pressure tend to include those reliant on the natural environment, cities, areas on the main tourist routes and National Parks. Those not under pressure includes those wishing to develop tourism currently with low visitor numbers and those off the beaten track. Interestingly, 73% local tourism area authorities at (ZAWA, NHCC) respondents (Southern Province region) perceived Livingstone greater areas to be under pressure compared with 29% at central government department at MTENR. Explanations for the perceived higher pressure on the South include respondents' personal experiences in conjunction with often heavy concentrations of packaged tourism and adventure tourism utilising the physical and natural environment.

5.6. Linkages and Synergies Planning (MTENR/ ZAWA/NHCC/ZEMA)

Under Zambian laws consents are required for all tourism developments. Consents are issued by multiple central governmental departments, regional and local authorities and communities depending on the scope of the consent sought [60]. Ascertaining accurate data on tourism related resource consent applications is highly problematic. While many respondents were able to give precise numbers in relation to resource consent applications and refusals, a significant 13 respondents were not able to provide the data. The main reason given for this is that tourism is not always isolated as a key variable in the database recording process for tourism enterprise concession consent applications. Some developments are not primarily designed for tourism purposes but may produce a tourism spin-off, e.g. development of a winery. In other cases, databases are not set up to be readily searched, data is not feed into system as "tourism", but as "commercial activity" and in several cases, the detail of activity or data is not even kept. This seems to indicate an inherent problem in the data management of tourism enterprise concession consent applications with regard to tourism, and a technical inability to retrieve useful information that can inform tourism planning at local, regional and national strategic levels. Acknowledging the limitations of the data, the following results give a broad indication of the workings of the ZAWA process in relation to tourism development within local communities.

Twenty four respondents representing 56% had dealt with tourism enterprise concession consent applications since 2010. The highest number of applications dealt with by one authority was 40. Ten authorities had dealt with between 1 and 10 applications, six between 11 and 20 applications, five between 21 and 30 applications, and three had dealt with 31 or more applications. While the largest number of applications were dealt with by District Councils, participants in the process held the view that all of these tourism programmes are developed by a monoactor form of centralised administration, generally overlooking the knowledge, skills and goals of local tourism organisations, both public and private, 50% of the CBNRM accounted for 37% of tourism enterprise concession which still needed ZAWA approval, indicating a substantial number of applications within a small number of local communities taking the lead role in resource management. Some 76% of tourism concessions submitted were made to the now well established Mukuni Community Development Trust. The trust has established local progressive leadership and used African Wildlife Foundation (AWF) assistance to develop twelve lower level area boards, which is an encouraging result suggesting that local CBNRMs playing a role in receiving enterprise concession related to tourism might have a strategic vision of how tourism should develop in their locality. Importantly, most of the communities receiving large numbers of tourism enterprise concession did have some form of policy guidelines, although two respondents received more than 25 applications did not.

Further, 24% of enterprise concessions were submitted to a start up Sekute Community Development Trust without a tourism guidelines or policy. There is no particular pattern of number of tourism enterprise concession received and the visitor numbers in remote CBNRM areas, with the largest numbers of applications 8 of respondents at ZAWA with over 25 applications in a variety of rural and urban environments, representing those areas that are already important tourism hubs (3 of the 8) and those encouraging the development of a tourism economy (5 of the 8). 3 of respondents at ZAWA received no applications, all of which are in insignificant under developed tourism areas: two not on tourist routes and one within a provincial city environment. One might expect a relationship between those ZAWA provincial offices reporting a large number of tourism enterprise concessions and those reporting that they perceived their area to be under pressure from tourism but this was not the case. 8 ZAWA respondents reported 25 or more applications, 5 respondents stated that their area was not under pressure from increasing tourism. In fact of the 19 of respondents at ZAWA that reported their area to be under pressure, 9 respondents were not able to extract numbers relating to tourism pressure, one ZAWA respondent had never handled tourism enterprise concession applications. A further four respondents received fewer than 10 tourism enterprise concessions, suggesting that it is not necessarily new developments that are creating tourism pressures. Indeed, one might say that applications made under the post [12,50] strategy are perhaps less problematic than existing developments that already generate significant demand.

The notable major challenges identified in the survey include that of poor understanding of what is required in the application for tourism developments activities. Eight respondents 24% of those that had experienced difficulties with applications stated that applications are often presented with incomplete information and a further eight respondents 24% identified lack of understanding and requirements for the tourism establishments under [35,50] process to be a reason why problems are experienced in the application procedure. However, as one respondent commented, early contact with the authorities is important for the process to run smoothly for the applicant: "it is not as bad as they initially think". Similarly, a further difficulty in applications is a lack of consideration of impacts of developments (18%). However, 21% of those that had dealt with ZAWA applications had not experienced any difficulties. As one respondent commented, "ZAWA act is there to protect the environment if a tourism developer follows carefully with ZAWA planning and guidelines/tourism experts, then things appear to go relatively smoothly. Communication between all parties is the key ingredient". The relationship between tourism development, sustainability and the Zambia Tourism Policy and ZAWA act towards tourism, as stated by a respondent: "at the moment the

MTENR Tourism Policy of 1999 and ZAWA act of 1998 deals with the sustainability of tourism on a case by case basis, however, at a strategic level the sustainability of tourism is not grappled with, due to outdated policy, legislation and planning". It is also apparent that the Tourism Policy/ZAWA act does not necessarily assure a sustainable approach to tourism planning outside of the particular development under consideration. For example, one respondent noted that: "results from Zambia Tourism Policy and wildlife authority" has been beset with negative administration, transparency and accountability issues". While it is unclear to what extent planning officers work with developers to ensure resource consents are granted, the general premise that there are few outright refusals begs the question as to whether the ZAWA process is rigorous in controlling the negative impacts of tourism in areas under pressure from increased visitor numbers. One respondent commented that "the two institutions MTENR and ZAWA are not a detractor to tourism development", which may or may not be a good thing.

6. IMPLICATIONS

It is clear from the survey findings that the dual role of MTENR/ZAWA in performing a regulatory planning function and promoting tourism raises issues about potential conflicts of interest in applying the Tourism Policy, ZEMA and ZAWA acts while considering the economic development of a locale. This debate is an old one environment versus economics, but in a sustainable development context the need to protect environmental resources to ensure future economic stability is mandatory. It is clear from observations of local communities and authorities at local and regional offices have engaged more actively with the tourism sector through the development of tourism plans and policies. In order to have sustainable development as a national policy direction as reflected in policy developments in the revised sixth national development plans (SNDP: 2011-2013).

6.1. Roles and Integrations of Sustainable Strategies

The integration of Sustainable Development (SD) Strategy in Zambia's National Development Plans' legal framework, various legislation in support of SD developed such as, ZEMA act (2011) to address impacts through strategy preparation is encouraging. However, due inability of local community authorities to benefit directly from the limited resources, especially those with a small population base and limited ability to raise revenue through rates, providing infrastructure, promoting tourism growth and managing impacts are a financial burden on tight budgets from central government: this emerges as a clear theme in the survey. New legislation currently under consideration to minimise waste provides a refund to communities, this is one example of where finding ways to compensate local communities and ratepayers for the use of local services is clearly a challenge and for many councils in Zambia and, indeed, worldwide, juggling the economic costs and benefits of tourism and justifying the outcomes to ratepayers remains problematic.

6.2. Delayed Decentralization Tourism Planning

This study shows that local authorities understand the roles of the MTENR/ZAWA with regard to sustainable tourism, focusing on the effects of tourism activity within their area. "Looking at the bigger picture, one of the criticisms of haphazard sort of implementation due to silo national level planning" [79]. As such, while the intentions of ZAWA in preventing undesirable developments are laudable, the cumulative effects of a number of seemingly innocent, less damaging developments might be equally detrimental. Only one respondent specifically drew attention to this issue, but that does not detract from the importance of the point indeed it might be questioned whether planning officers are sufficiently aware of the dangers posed by this breach within ZAWA framework. Similarly, the focus of ZAWA on effects of activities, while well intentioned, could result in significant economic sectors, like tourism, not been adequately and proactively planned for. Somewhat worryingly, this might be reflected in the lack of response from regional agencies, who do not appear to take tourism as a specific concern under their remit, although are clearly concerned by the effects of tourism such as waste. The inherent difficulties of extracting tourism related projects from MTENR/ZAWA databases held by local tourism authorities appears to be an issue in understanding the implications of the ZAWA for tourism and the extent to which projects are acceptable in the local planning decisionmaking process. Quite clearly, this reflects the inadequacies within data management and retrieval, but also indicates a systemic challenge for the core workings of ZAWA, which by its nature is not concerned with specific industry sectors but with the effects of activities. While the key focus on natural resources provides a valuable framework for the development of appropriate policy and decision-making frameworks, the ability to understand the scope and scale of tourism-related developments is essential particularly given the ambitions of the proposed national tourism strategy (SNDP). Worthy of note is that of the 13 local authorities that were unable to retrieve tourism-related data due to technical problems of record keeping and searching were: four of the eight central government departments (MTENR) stating that their areas were under pressure from increasing tourism; further, two of Zambia's new prime Lusaka circuit very significant local tourist locations; and, further again, three other well-known Livingstone tourism areas. These omissions from knowledge at a planning level indicate the potential to not fully understand the rate of tourism growth from a supply perspective and the cumulative effects of tourism development linked with local aspirations within the confines of long term national development planning (plans) and a budgeting system that is partially decentralised.

7. CONCLUSIONS

The continuing limited involvement of local communities and regional government authorities in tourism planning and development of sustainable tourism approaches existed, given "the continued conspicuously absence of documented national development planning policy and fragmented legislation framework of sustainability in Zambia's national strategies"[60]. With the role of tourism in economic development established and recognised in statutory plans, sustainability now underpins sectorial policy framework for tourism in Zambia, and the landmark steps taken to develop and review national aspirations for tourism development will represent a step forward in establishing a clear remit for local government in planning for tourism. The extent to which this is rhetoric rather than reality is questionable, given the somewhat mixed results in the survey of local government agencies reported in this study. For Zambia a country emerging from a history of centralised economic planning, this question becomes even more vexed quite clearly, a range of pressures continue to affect local areas and the challenges that face many local communities in trying to manage the effects of tourism on environmental resources are as pressing as ever. A national tourism plan will enable local authorities and councils to evolve futures that befit environmental resource opportunities and constraints, community aspirations and local budgets. While tourism is mainly a private sector industry in Zambia, the public sector adopts a dual role as the gatekeeper of tourism developments through planning control, while promoting economic development opportunities through tourism.

As such, while councils have become the arbiters of sustainable tourism through their role in implementing the Zambia Tourism Policy, the appeal of developing the local economy places them in a dichotomous position. While much of this discussion sounds positive, there is still a major gap between strategy and implementation in the evolution towards Zambia as a sustainable destination. While sustainability is now one of the cornerstones under tourism strategy review, much of this lies at a national strategic level and remains as a philosophical stance. Evidence suggests that problems created by tourism pressures do exist and some of these are difficult to deal with given the poor linkages and synergies within the various tiers of government that undertake planning with limited budgets at local government. Pressure at key tourist hotspots and with certain tourism related activities are recognised and with the continuing growth in tourist numbers forecasted, the effects of tourism have the potential to change the nature of the tourist experience and the very foundations on which Zambian tourism is built. The existing problems of geographic concentration of tourism activity will only worsen, exacerbating the pressures on local authorities.

As argued by [64] "policy at a national level that assists local areas in dealing with volumes and the distribution of tourists in a more methodical manner". With reference to [80-84] by enabling more proactive public sector approach to tourism planning, steps towards understanding the dynamics of tourism in Zambia made by the Ministry of Tourism Environment Natural Resource under the Zambia Wildlife Authority by establishing a strategic tourism development model. Given that local government agencies are politically weak, of well-recorded and entrenched patterns of corruption and patronage built around land and planning decisions, this call by planners has a greater degree of cogency as observed by [60,80,85] argue that, "those destinations, localities and nations that prepare to put into practice good detailed policies and strategic plans will reap the benefits for sustaining their tourism products in the future", a cornerstone of Zambia's tourism strategy. Further research and steps would help local Zambian destinations to ensure ZAWA achieves the goals and principles enshrined in the original legislation. Without a more concerted attempt to challenge pro-development policy, Zambia is likely to lose pace in terms of competitive advantage as a clean, green and sustainable tourism destination.

ACKNOWLEDGEMENTS

The authors wish to express their gratitude to Alexandra Thorer & Roy Clarke (kalakikorner) for the assistance in making suggestions, corrections and to the would be referees for their valuable and helpful comments, which have improved the quality of the paper. Also, we extend our thanks to The Copperbelt & Tongji University, especially the Department Landscape studies.

REFERENCES

1. A. Hansen, "The Ecotourism Industry and the Sustainable Tourism Eco-Certification Program (STEP)," International Relations and Pacific Studies, University of California, San Diego, 2013. irps.ucsd.edu/assets/021/8446.pdf
2. C. Wu, "Sustainable Development Conceptual Framework in Tourism Industry Context in Taiwan: Resource Based View," Conference of the International Journal of Arts and Science, Vol. 2, No. 1, 2009, pp. 1-11.
3. C. A. Gunn, "Tourism Planning," Routledge, London, 2002.
4. B. Ahn, B. Lee and C. S. Shafer, "Operationalising Sustainability in Regional Tourism Planning: An Application of the Limits of Acceptable

Change Framework," Tourism Management, Vol. 23, No. 1, 2002, pp. 1-15.

5. UNEP, "Eco-Tourism in Wider Caribbean Region: AN Assessment," Technical Report No. 31, Caribbean Environment Programme Technical, UNEP, Kingston, 1994.

6. UNEP, "Environmental Codes of Conduct for Tourism," United Nations, New York, 1995.

7. World Tourism Organisation (WTO), "Guide for Local Authorities on Developing Tourism," WTO, Madrid, 1999.

8. C. Tosun, "Challenges of Sustainable Tourism Development in the Developing World: The Case of Turkey," Tourism Management, Vol. 22, No. 3, 2001, pp. 289-303.

9. K. Andriotis, "Tourism Planning and Development in Crete, Recent Tourism Policies and Their Efficacy," Journal of Sustainable Tourism, Vol. 9, No. 4, 2001, pp. 298-3016.

10. V. M. Waligo, J. Clarke and R. Hawkins, "Stakeholder Involvement in the Implementation of Sustainable Tourism," Journal Tourism Management, Vol. 36, 2013, pp. 342-353.

11. UNEP, "The Role of Local Authorities in Sustainable Tourism. Tourism and Local Agenda, 21," United Nations, Rio, 2003.

12. Zambia Wildlife Authority (ZAWA), "Annual Report," Government of the Ministry of Tourism, Environment and Natural Resources, Republic of Zambia Kwacha House, 2007.

13. V. Teye, "Tourism Development in Zambia: Some Physical and Environmental Considerations," Invited Chapter in T. V. Singh and H. L. Theuns, Eds., Towards Appropriate Tourism: The Case of Developing Countries. Peter Lang (European University Studies), Frankfurt, (Revised and Expanded Version of "Geographical Factors Affecting Tourism in Zambia," Annuals of Tourism Research, Vol. 15, No. 4, 1988, pp. 487-509).

14. C. A. Gunn, "Prospects for Tourism Planning: Issues and Concerns A," The Journal of Tourism Studies, Vol. 15, No. 1, 2004, pp. 3-7.

15. Government of the Republic Of Zambia, "Tourism Sector Profile," Zambia Development Agency (ZDA), Lusaka 2013.

16. Zambia Tourism Board, "2010 Visitors' Arrivals Analytical Report," Ministry of Tourism, Environment and Natural Resources, Government of the Republic of Zambia, Kwacha House, 2011.

17. World Bank, "Zambia Economic and Poverty Impact of Nature-Based Tourism," Report No. 43373-ZM, Economic and Sector Work Africa Region 2007.

18. World Tourism Organisation (WTO), "Tourism Highlights 2012" Calle del Capitán Haya, Madrid, 2012.

19. United Nations Environment Programme and World Tourism Organisation, "Tourism in the Green Economy," Background Report, UNWTO, Madrid, 2012.

20. USAID and Sustainable Tourism, "Meeting Development Objectives, USAID's Portfolio in Sustainable Tourism," US Agency for International Development Washington DC, 2005.

21. J. E. Mbaiwa, "Tourism Development, Rural Livelihoods, and Conservation in the Okavango Delta, Botswana, A Doctor of Philosophy Dissertation," The Office of Graduate Studies of Texas A&M University College Station, 2008.

22. W. Jamieson, "Guidelines on Integrated Planning for Sustainable Tourism Development," Economic and Social Commission for Asia and the Pacific, 1999.

23. S. Chiutsi, M. Mukoroverwa, P. Karigambe and B. K. Mudzengi, "The Theory and Practice of Ecotourism in Southern Africa," Journal of Hospitality Management and Tourism, Vol. 2, No. 2, 2011, pp. 14-21.

24. PATA, "Endemic Tourism: A Profitable Industry in a Sustainable Environment," Kings Cross, 1992.

25. WTO, WTTC, Earth Council, "Agenda 21 for the Travel and Tourism Industry: Towards Environmentally Sustainable Development," WTTC, London, 1995.

26. C. Hall, "Historical Antecedents on Sustainable Development: New Labels on Old Bottles?" In: C. M. Hall and A. A. Law, Eds., Sustainable Tourism: A Geographical Perspective, Longman London, 1998, pp. 13-24.

27. UNESCAP, "Managing Sustainable Tourism Development: Tourism Review No. 22," United Nations, New York, 2013. http://www.unescap.org/ttdw/Publications/TPTS_pubs/Toreview_No22_2141.pdf

28. M. Lozano-Oyola, F. J. Blancasa, M. González and R. Caballero, "Sustainable Tourism Indicators as Planning Tools in Cultural Destinations," Journal Ecological Indicators, Vol. 18, 2012, pp. 659-675.

29. International Union for Conservation of Nature (IUCN), "Directory of Afro-Tropical Protected Areas," IUCN, Gland, 1987.

30. S. Metcalfe, "Landscape Conservation and Land Tenure in Zambia: Community Trusts in the Kazungula Heartland," African Wildlife Foundation, Harare Zimbabwe, 2005.

31. V. R. Nyirenda, S. Liwena and H. Kaumba Chaka, "Atlas of the Natonal Parks of Zambia," New Horizon Printing, Lusaka, 2008.

32. Central Statistic Office, "National Accounts Statistical Bulletin," Republic of Zambia 2010.

33. K. T. Taylor and C. T. Banda, "Tourism Development Potential of the Northern Province of Zambia," American Journal of Tourism Management, Vol. 2, No. 1A, 2013, pp. 10-25.

34. Policy Monitoring and Research Centre (PMRC), "Tourism and Wealth Series—Unlocking the Potential of the Tourism Sector to Support Economic Diversification and BroadBased Wealth," Lusaka, 2013.

35. Zambia Wildlife Authority (ZAWA), "Turn-Around Strategy (2011-2016) for the Zambia," 2011.

36. Zambia Tourism Board, "2009 Visitors' Arrivals Analytical Report," Ministry of Tourism, Environment and Natural Resources, Government of the Republic of Zambia, 2010.

37. P. P. Wong, "Coastal Tourism Development in Southeast Asia: Relevance and Lessons for Coastal Zone Management," Ocean & Coastal Management, Vol. 38, No. 2, 1998, pp. 89-109.

38. B. Hatyoka, "UNWTO Offers Great Benefits to Livingstone," Times of Zambia, Business Insights: Essentials. Web. 23 October 2013. UNWTO Offers Great Benefits to Livingstone, 28 August 2013.

39. The Government of the Republic of Zambia, "Zambia National Report," The United Nations Conference on Sustainable Development (Rio+20), Government Printer, Lusaka, 2012.

40. "Reposition Zambia on Tourism Map," Times of Zambia 5 December 2012. Business Insights: Essentials. Web. Retrieved on 23 October 2013.

41. Zambia Wildlife Authority (ZAWA), "The General Management Plan (GMP) for Mosi-oa-Tunya National Park (MoNP)," 1998.

42. Zambia Wildlife Authority, "Five-Year Strategic Plan 2003- 2007," Volume Implementation Plan, Chilanga, 2002.

43. International Union for Conservation of Nature (IUCN), "Biodiversity in Sub-Saharan Africa and Its Islands," IUCN, Gland, 1990.

44. E. Chidumayo, F. Lumbwe, K. Mbata and J. Munyandorero, "Review of Baseline Status of Critical Species Habitats in the Mosi oa Tunya National Park," Department of Biological, University of Zambia, Lusaka, 2003.

45. "Tourism Contributes to Environmental Degradation-President Sata," Lusaka Times, 29 May 2012. Home News: Web. 2012. http://www.lusakatimes.com/2012/05/29/tourism-constributes-environemtal-degradationpresident-sata/

46. B. Liu and F. M. Mwanza, "Questionnaire Website, Sustainable Tourism Development in the Case of Livingstone Zambia," 2012. http://floydmwanza.achievestdzambia.questionpro.com/

47. D. Dredge, "Policy Networks and the Local Organisation of Tourism," Tourism Management, Vol. 27, No. 2, 2006, pp. 269-280.

48. T. Holding, "Environmentally Sustainable Tourism Strategic Plan 2009-2012," Tourism Victoria Victoria, 2009.

49. Government of the Republic of Zambia, "The Tourism Policy for Zambia," Government Printer, Lusaka, 1999.

50. Zambian Environmental Management Agency ZEMA, "Environmental Management Act (EMA)," No. 12. Government Printer, Lusaka, 2011.

51. Policy Monitoring and Research Centre (PMRC), "Tourism Series, Tourism: Supporting Sustainable Development, Income and Job Creation, Zambia," 2013. www.pmrczambia.org.

52. Government of the Republic of Zambia (GRZ), "Fifth National Development Plan: Sustained Economic Growth and Poverty Reduction," Government Printer, Lusaka, 2006.

53. Government of the Republic of Zambia (GRZ), "Sixth National Development Plan: Sustained Economic Growth and Poverty Reduction," Government Printer, Lusaka, 2011, p. 143.

54. Grant Thornton Kessle Feinstein, "Study for Development and Promotion of Tourism in Zambia," 2003.

55. Government of the Republic of Zambia, "Formulation of the National Adaptation Programme of Action on Climate Change (Final Report)," Ministry Of Tourism, Environment and Natural Resources, Lusaka, 2007.

56. Government of the Republic of Zambia, "The National Environmental Action Plan," Ministry of Environment & Natural Resources, Lusaka, 1994.

57. UNEP, "Biodiversity Planning Support Programme, Guide to Best Practices for Sectoral Integration: Integrating Biodiversity into the Tourism Sector," United Nations, New York, 2001.

58. UNEP, "Report on Industry and Sustainable Tourism for the 7th Session of the CSD Tourism and Environment Protection," United Nations, New York, 1999.

59. UNCED, "Agenda 21," United Nations Conference on Environment and Development, Rio de Janeiro, 1992.

60. S. Berrisford, "Revising Spatial Planning Legislation in Zambia: A Case Study," Urban Forum, Vol. 22, No. 3, 2011, pp. 229-245.

61. Zambia Wildlife Authority-ZAWA, "Final Business Plan," Government of the Ministry of Tourism, Environment and Natural Resources, Republic of Zambia, 2006.

62. Zambia Land Alliance (ZLA), "Land Policy Options for Development and Poverty Reduction Civil Society Views for Pro-Poor Land Policies and Laws in Zambia," Lusaka, 2008.

63. K. Andriotis, "A Framework for the Tourism Planning Process," In: A. Raj, Ed., Sustainability, Profitability and Successful Tourism, Kanishka Publishers, New Delhi, 2007.

64. J. Connell, S. Page and T. A. Bentley, "Towards Sustainable Tourism Planning in New Zealand: Monitoring Local Government Planning under the Resource Management Act," Tourism Management, Vol. 30, No. 6, 2009, pp. 867-877.

65. K. Angelevska-Najdeska and G. Rakicevik, "Planning of Sustainable Tourism Development," Procedia-Social and Behavioral Sciences, Vol. 44, 2012, pp. 210-220.

66. K. Kelley, B. Clark, V. Brown and J. Sitzia, "Methodology Matters: Good Practice in the Conduct and Reporting of Survey Research," International Journal for Quality in Health Care, Vol. 15, No. 3, 2003, pp. 261-266.

67. E. R. Babbie, "Survey Research Methods," Wadsworth, Belmont, 1990.

68. J. E. Bartlett II, J. W. Kotrlik and C. C. Higgins, "Organizational Research: Determining Appropriate Sample Size in Survey Research," Information Technology, Learning, and Performance Journal, Vol. 19,

No. 1, 2001, pp. 43- 50. http://www.osra.org/itlpj/bartlettkotrlikhiggins.pdf.

69. A. Attia, "Planning for Sustainable Tourism Development: An Investigation into Implementing Tourism Policy in the Nwc of Egypt," Thesis Degree of Doctor of Philosophy, Development Planning Unit, The Bartlett School Of Architecture and Planning University College London, University Of London, London, 1999.

70. E. G. Helmy, "Towards Sustainable Tourism Development Planning: The Case of Egypt," A Thesis Submitted in Partial Fulfillment of Doctor of Philosophy, Bournemouth University, Bournemouth, 1999.

71. D. Dierckx, "How to Estimate Your Population and Survey Sample Size," 2013. http://www.checkmarket.com/2013/02/how-to-estimate-your-population-and-survey-sample-size/0.

72. "The Poverty Reduction Strategy Paper (PRSP)," Ministry of Finance and National Planning, Government of the Republic of Zambia, 2002.

73. Grant Thornton Associates Limited, "Audit of Zambia Wildlife Authority's (ZAWA's) Current Capabilities and Development of Project Proposals for Institutional Capacity Development over a Five Year Period (2004-2008)," Lusaka, 2004.

74. Government of the Republic of Zambia, "Visitors Arrivals Analyticl Report," Produced By: Ministry of Tourism, Environment and Natural, Lusaka, 2009.

75. World Bank, "Zambia-Support to Economic Expansion and Diversification (SEED): Tourism," Republic of Zambia Ministry of Tourism, Environment and Natural Resources Securing the Environment for Economic Development (SEED) NW, Washington D.C., 2001. http://www.worldbank.org/infoshop

76. Government of the Republic of Zambia, "Mosi-oa-Tunya National Park General Management Plan," ZAWA, Lusaka, 2000.

77. GEF-UNDP-Anchor Environment Consultants, "2001-2006 Economic Analysis and Feasibility Study for Financing Zamibia's Protected Areas," GEF-UNDP-Government of Zambia, 2001.

78. European Development Fund (EDF), "National Parks and Wildlife Service Project: Draft Master Plan," Transtec, Milan, 1998.

79. C. K. Chunga, "Integration of Sustainable Development in Zambia's. National Development Plans," Lessons Learnt &. Recommendations, National Planning Department, Ministry of Finance, 2012.

80. D. L. Edgell, M. D. Allen, G. Smith and J. R. Swanson, "Tourism Policy and Planning: Yesterday, Today and Tomorrow," Butterworth-Heinemann, Oxford, 2008.

81. N. Blaikie, "Designing Social Research, An Introduction to Qualitative Research," Blackwell, Oxford, 2000.

82. P. A. DeGeorges and B. K. Reilly, "Sustainability, the Realities of Community Based Natural Resource Management and Biodiversity Conservation in Sub-Saharan Africa," Economic Growth and Sustainable Wildlife Management, Vol. 1, No. 3, 2009, pp. 734-788.

83. R. Dodds, "Sustainable Tourism: A Hope or a Necessity? The case of Tofino, British Columbia," Canada Journal of Sustainable Development, Vol. 5, No. 5, 2012, pp. 54-64.

84. A. J. Dougill and M. S. Reed, "Framework for Community-Based Rangeland Degradation Assessment for the Kalahari, Botswana," RGS-IBG Annual Conference DARG Session on Sustainable Resource Use: Critical Issues in Developing Areas, The Royal Geographical Society, Kensington Gore, 2003, p. 23.

85. L. Dwyer, D. Edwards, N. Mistilis, C. Roman, N. Scott and C. Cooper, "Megatrends Underpinning Tourism to 2020: Analysis Of Key Drivers For Change," CRC for Sustainable Tourism Pty Ltd., Gold Coast, 2008.

Chapter 2

Recognizing Emotion and its Postcolonial Potentialities: Discomfort and Shame in a Tourism Encounter in Turkey

Hazel Tucker

Department of Tourism, University of Otago, Dunedin, New Zealand

ABSTRACT

This article addresses the fragile and potentially problematic nature of the meeting of tourists and local 'hosts' by shifting the discussion away from the authentic/fake binary and focusing instead on emotion in the worldmaking tourism encounter. This is done through interrogation of one particular encounter which took place in Göreme, central Turkey, between a local woman, a German couple and me. The problematic nature of the encounter exposed the point that tourism encounters are not reducible to questions of discourse alone and that, if we are to understand tourism encounters more fully, it is necessary to examine closely their emotional and bodily dimensions. Moreover, it is argued that recognizing and acknowledging emotion, and particularly shame, presents the postcolonial potentialities of tourism and Tourism Studies in that the discomfort of shame can produce a positive disruption of the otherwise inherently colonial relationship between tourist and other. In turn, reflexive interrogation of my own discomfort prompts discussion of researchers' ability to know what is going on in the production of local worlds and, indeed, to be aware of their own often-powerful role in the worldmaking function of Tourism Studies.

Keywords: Encounter, emotion, shame, postcolonial, reflexivity, the declarative role of tourism, rencontre, émotion, honte, postcolonial, réflexivité, le rôle déclaratif du tourisme, Begegnung, Gefühl, Scham, postkolonial, Reflexivität, die erklärende Rolle des Tourismus

INTRODUCTION

Late one morning I was sitting outside in the street chatting to three local women. They all lived close by the cave-house that my family and I were renting for the duration of our 'sabbatical' stay in Turkey. Upon seeing a tourist couple walking by one of the women, my next door neighbour Emine, called out to them, 'Come, church!' The tourists didn't understand but came over to see what she was saying. I quickly gathered that the tourists were German but, assuming they would speak some English, I jumped in to help and translated my neighbour's question: 'She's asking if you want to see a church'. They said they did, so we all followed Emine. I hadn't actually been inside this church even though it was situated right on the short narrow street where I was living. We went through a doorway of an empty ruined cave house and inside was the dark church dug out of the rock. It was full of rubble and rubbish and the German couple didn't seem very impressed. Emine looked a bit disappointed in that. She told me, in Turkish, that some of the tourists she shows in here go 'Ooh, aah' when they see it. The German man asked how old the church was and I translated the question to Emine. I translated the reply that it was around a thousand years old.

Emine then asked if they wanted to see her cave house. They said they did and I went too, at the request of Emine, in order to translate. Through Emine's front gate we entered a courtyard that was open on one side giving a spectacular view over the village and the Göreme valley beyond. The German couple stood and marvelled at the view. Emine then led us into a sort of make-shift cave kitchen and invited us to sit down on stools around a small table. She placed a plate of grapes onto the table, saying that they had come from her sister's garden. A conversation about family ensued with me translating. Emine asked the Germans if they had any children and they replied that they had three. Emine explained that she got divorced when her daughter was three years old and that her parent's said that she could come and live in the family's old cave house. Her father and mother had died twelve and fifteen years ago respectively. Emine asked the German couple if they liked Göreme and they said that they did. Emine said that when she's coming home from town there are sometimes tourists on the bus also and she notices them being taken aback by the view as they come down into the Göreme valley. She said that the Göreme people were used to the views.

'Doing tourism', according to Crouch (2005), is a process of seductive encounter in which expectations, experiences and desires are continuously negotiated. Tourism desire has thus been described as similar to a love affair (Lengkeek 2002), as a flirtatious encounter (Crouch 2005). The flip side of this, however, is that the tourist encounter is always potentially problematic, being at the same time both full of promise but also having the potential to not meet that promise. Many previous analyses of such a threat to a positive tourist experience have 'adopted a censorious moral tone condemning the corruption and imminent collapse of tradition in relation to "other" people and culture' (Meethan 2001: 165) as the probable cause of the tourist experience not meeting expectations. In this paper, I want to show how we can shift our discussions of tourism encounters away from the authenticity/fake binary by focusing on the emotional dimensions of the encounter and by acknowledging that *embodied* tourism encounters 'are always open to individual reinterpretation and the potential to be disrupted by the fleshiness of the body' (Waitt *et al.* 2007: 252).

The paper has begun and will continue by describing one particular encounter that took place in Göreme, in the Cappadocia region of central Turkey, between a local woman, a German couple and me. I am a female New Zealand-based researcher with previous ethnographic research experience in the region, that research broadly looking at tourism relationships and the socio-cultural implications of tourism for the Göreme community. The awkwardness I felt as the specific encounter upon which this article is based unfolded has highlighted not only the problematic nature of tourism encounters, but also the way in which I, in my own ethnographic work, am always fully and bodily implicated in the creation of tourism knowledge. Prompted by Massumi's (2002: 18) argument that 'experimentation activates detail', my experimentation in this article through interrogation of the discomfort I felt in the encounter draws out detail concerning my own positioning in this touristic milieu. This detail relates to the point made by Ateljevic and Hall (2007: 139) that tourism researchers must fully accept and acknowledge that 'they are always an interlocutor charged with the final act of interpretation and representation of knowledge'. In the words of Hollinshead (2004: 32), also, they 'must not be blind to the sorts of textual authority which runs through them as that very tourism textuality helps render or make the world'.

Furthermore, my reflexive engagement in the particular encounter upon which this article is based has made me recognize the tensions that arise in my attempts to combine into one my 'doing ethnography' with the toured Other and my 'doing ethnography' with the tourist. Studying both tourists and toured is a prime example of the rich anthropological studies of the contemporary that Holmes and Marcus (2005: 1103) describe, which incorporate 'both communities of often elite discourse and communities of often subaltern subjects'. However, as Holmes and Marcus (2005: 1103) also point out, such studies inevitably render an 'uneven distribution or depth of knowing in ethnography'. Whilst the suggestion is not unfamiliar that 'doing ethnography' is not too far removed from 'doing tourism' (Crick 1995), in my previous ethnographic work in Göreme (see, particularly, Tucker 2003) I have tended to treat the tourist as if she or he were a distanced 'other' in relation to me, just as I have done with the toured Göreme 'other'. In this article, through discussion of the emotional-affective dimensions of the encounter, I have become more aware of my own alignment, not only with the (colonizing) tourist, but also with what Hollinshead (2004) refers to as the 'declarative' value of tourism and the 'textual authority' of Tourism Studies. Rather than inhibiting my ethnographic work, however, this awareness has led to an understanding of the productive and *anti*-colonial role that reflexivity and recognition of emotion, and especially shame, can play in both the tourism encounter and Tourism Studies.

My focus here on awkwardness and emotion in the encounter leads me to make two separate but intertwined arguments. The first is that whilst the tourism encounter and the creation of tourism knowledge 'cannot be separated from notions of power/ideology and micro-politics of cultural negotiation' (Ateljevic and Hall 2007: 139), tourism encounters are not reducible to questions of discourse alone. It is therefore necessary to examine closely the emotional and affective or bodily dimensions of the encounter if we are to begin to understand the 'subjective seductions' of tourism in Crouch's (2005: 26) words. Some detailed consideration of the body in tourism has already occurred in Tourism Studies (for example, Veijola and Jokinen 1994; Crouch 2002; Andrews 2006; Veijola and Valtonen 2007). However, discussions of the embodied dimensions of tourism, including discussions of gender in tourism, have tended largely to refer to the body in rather general and abstract terms. As such, there remains a lack of understanding of how the physical body can disrupt the flow of discourse in tourism. This leads to the second contention in this paper which is that recognizing and acknowledging emotion, and in particular shame, opens up postcolonial potentialities in tourism and Tourism Studies. I base this point upon Slemon's (1991: 3) definition of postcolonial:

> Definitions of the postcolonial, of course, vary widely, but ... the concept proves most useful not when it is used synonymously with a post-independence historical period in once-colonised nations, but rather when it locates a specifically anti- or *post*-colonial *discursive* purchase in culture.

With tourism clearly very 'closely entwined with the imperial project and colonialism' (Pritchard and Morgan 2007: 21), it is important to examine in detail the moments in which the colonizing worldmaking practices of tourism are disturbed by the body and emotion. To do so in this article, I focus in-depth on the discomfort, both felt and apparent, in this one tourism encounter in Göreme by drawing in particular on Probyn's (2005: xviii) illumination of the ways in which shame can be 'productive in how it makes us think again about bodies, societies, and human interaction'.

CONTINUING THE ENCOUNTER

As we sat in the cave kitchen eating grapes, the conversation, as well as my position, got difficult for a while. Emine started talking at length about the Göreme mayors and how some had done a lot for the village and some had not. I translated when I got a chance but was aware that it was too in-depth and therefore potentially boring for the Germans. The Germans tried to make conversation with me too by telling me, upon my telling them that I had moved to live and work in New Zealand 5 years ago, where they had travelled in Australia. Emine couldn't understand and she countered this with a long rant which I could barely follow about local politics. The German couple began to look very bored. I wondered what Emine was doing. Why wasn't she entertaining them a little more congenially? Perhaps she didn't know how to. I felt uncomfortable playing the rope in this conversational tug of war, so I jumped up and suggested to Emine that the tourists might like to see more of the cave-rooms of the house.

We went into a beautiful arched room, the room normally used for receiving guests. In there, Emine immediately emptied the contents of a plastic bag onto the floor. She invited the German woman to sit down and have a look at the pile of headscarves and knitted socks now on the floor. The man remained standing and gazed out of the window. The couple suddenly became quieter, perhaps sensing now that the whole thing had just been about selling. I felt embarrassed. I knew that Emine would bring out some handicrafts at some point in the hope of selling them but she had done so entirely without grace and had not, I felt, yet given the tourist couple an 'experience' to draw them in and to balance the obligation to buy something; she had not succeeded in enchanting them yet. The German woman picked slowly through the pile and said to me 'Well, I suppose I have to buy one now'. I felt very uncomfortable and felt my face blush: maybe the Germans were disappointed in me now as well and thought that I was colluding with Emine and was going to get a cut of the profit for myself. I tried to make a middle ground for myself, between being a bit surprised at what Emine had done, but also showing that I had semi-expected this to happen. I felt compelled to try to rescue their experience of Emine and her house. I attempted to make the headscarves more interesting by asking Emine how she had done the different designs of the croched and beaded borders and then translating for the Germans. Emine then said that, because the borders had different levels of difficulty they varied in price and she listed the rather high prices of the different designs. The German woman, seemingly reluctantly, chose one to buy – a light blue one costing fifteen lira. She took out a twenty lira note and Emine went off to find change. Now that

Emine was out of the room the German woman turned me and, with her head slightly lowered, said 'Well, I suppose they have to make money somehow'. I agreed and reiterated the fact that Emine was divorced and had had to raise her daughter on her own. The Germans agreed but also laughed as if to show resignation to the fact that they had been duped by Emine.

When Emine returned with the change, compelled to now take over the situation and try to help the Germans have more of an experience in exchange for their purchase, I suggested that we look at more of the house, and particularly the food storage cave as I thought that that would be of interest to them. This cave had some large waist-high storage pots inside. The Germans asked what was in the pots and so I showed them and explained that that is how the villagers store their dry products such as bulgur wheat and dried chickpeas throughout the winter. I asked Emine if she had made the bulgur wheat herself and she replied that no, she had bought it. I tried to make it more interesting for the Germans by telling them that many of the village women make many such food items themselves from scratch, such as growing and processing bulgur wheat and pickling vegetable produce from their gardens using their own homemade vinegar. I told them how women form groups to make huge piles of flatbread which they also store in the caves for use through the winter. The Germans were becoming more interested and so I tried to engage Emine with their interest by asking her when she would be making her bread this year. She said that she didn't make bread because it makes her back ache. I didn't translate her reply to the tourists.

We went back out into the courtyard moving towards the exit of the house. The man took some photographs of the view and the house. He wanted a photo with Emine so I took one of them all with his camera. The man's wife then asked him to go back inside and take a picture of the cave-kitchen. He did that and meanwhile Emine showed the German woman how to wear her new headscarf Göreme-style. Just as we were leaving the house, Emine's teenage daughter arrived home. Emine pointed to her and said to the Germans 'this is my child' which I translated. The daughter protested and said 'not a child'. I translated that also. Everyone laughed and, with the German's already having told us that they had three grown-up children themselves, for a brief moment there was a sense of mutual understanding.

At the gate, the German couple and I said goodbye and thank you to Emine and walked off down the lane. The man said 'very interesting, it's not often you get to see inside people's houses'. The woman agreed with him: 'Yes, that was very interesting'. As we passed the ruined church the man said that maybe Emine should clean the church up a bit and put in some lights and then she could invite lots of tourists and charge them two lira entrance and then she would make much more money than with the headscarves. The man's wife and I laughed in agreement and then, as we reached the square by the mosque, the woman, still wearing the headscarf, looked at her reflection in the window of a parked car and gestured to acknowledge how silly she looked. I said 'Yes, you look very Turkish', laughing. While exchanging pleasantries and wishing each other enjoyable travels, she untied the headscarf to take it off and then we parted.

AWKWARDNESS AND DISCOMFORT IN THE ENCOUNTER

Emine, like some other women in the village, regularly invites passing tourists in to look at her cave-house in the hope of selling them handicraft items. Tourists' acceptance of such invitations is likely to stem from a desire to have a perceived 'authentically social' experience (Selwyn 1996; Tucker 2003). The fact that this encounter was serendipitous for the tourists meant that it held much promise right from the start and by the end of the encounter it seemed that the tourists felt that they had, after all, achieved something in getting 'backstage' into a villager's house. Emine brought out grapes rather than offer tea, since Turkish tea takes a long time to brew and so is usually offered only to proper guests and connotes the invitation to stay a long while. This encounter was never intended by any party to go on for too long. So the hospitality in offering the grapes, being a not-*too*-drawn-out opportunity for us all to share in the bodily consumption of some local produce, was received graciously by the tourists. Despite holding much promise, however, as it continued, the encounter cut an extremely fine line between satisfaction and discomfort for those involved, thus highlighting the fragility of the tourist encounter. In order to understand this awkwardness and discomfort, it is necessary to consider the positioning of those involved. I shall begin with the positioning of Emine in the tourism encounter through a brief discussion of gender relations in Göreme and how they relate to the worldmaking reach of tourism.

Despite urban women in Turkey enjoying increasing levels of equality and participation in the workforce (Sönmez 2001; Kandiyoti 2002), Göreme and the wider Cappadocia region remains a pocket of social ultra-conservatism more widely associated with the Islamic Middle East. My earlier case study research on tourism development in Göreme (Tucker 2003) outlined the powerful ideology of gender separation rendering tourism business and tourism space the male domain, whilst women, on the whole, were kept separate from tourism activity. Göreme women have generally only been able to access the tourism economy indirectly through the earnings of their husband or other male family members as they have remained separated from the business of tourism. In Göreme society it is shameful for a woman to wander around in public spaces because to do so is to put her in a position where she may come into contact with men other than those belonging to her household. If she remains always within the house (or the immediate neighbourhood) her virtue is protected. A woman's virtue is managed by her male relatives' giving or not giving of permission to enter public/male space, and it is also managed by gossip and the shame that gossip would bring. Also, because they have little experience of being in public spaces, women lack confidence in those spaces and feel uncomfortable.

Being separated from the spaces of tourism, the village women have largely continued to engage in garden-agriculture and other 'household' and reproductive duties. The women themselves and the cave-houses they inhabit have thus come to represent the 'traditional' in Göreme. Furthermore, tourism is derived from and perpetuates an aesthetic valuing of the rock cones and caves of Cappadocia, along with the cave-dwelling 'troglodytes' who inhabit them. During the past few years, increasing numbers of women have started capitalizing on this point by engaging in informal entrepreneurial activity by inviting tourists in to view their cave-houses and attempting to sell them handicrafts (Tucker 2007). The women engaging in this type of entrepreneurial activity acknowledge that they would not be able to work in tourism other than doing this sort of activity. Having their cave-house as an attraction and selling handicrafts from home therefore provides them with an opportunity to engage with the tourism economy and to earn some money of their own.

However, there is much gossip and criticism levelled at Emine and her neighbours concerning their tourism entrepreneurial activities, which is undoubtedly an example of the processes of boundary maintenance that occur whenever women engage in activities that are deemed inappropriate for their gender (Puwar 2004). Other Göreme women living in the same neighbourhood as Emine, women who themselves have enough financial security to not need to sell handicrafts to tourists, criticize the women who invite tourists into their homes, saying that their waiting on their doorstep for tourists to pass is akin to begging. The men, too, and particularly those men involved in more formal tourism business activities, criticize these women for their 'disturbing' the tourists. One man who owns a cave hotel situated in the street beyond Emine's house, for example, told me that he frequently hears from his guests of encounters similar to the one told above, with the tourists on one hand feeling pleased about having been into a 'real' village cave-house, but also feeling duped by the woman who invited them in when they realized she expected them to buy something from her. So, it is because the women remain limited in their ability to form successful interactions with tourists that the male tourism entrepreneurs consider the tourism activities of these women to be problematic . Simultaneously, because they are considered problematic such encounters remain hidden in the confines of domestic space, always shrouded in the women's discomfort and shame and always, therefore, slightly awkward.

Emine's awkwardness thus derives from her gendered positioning, upon which a particular local system of gossip and shame is a key mechanism of control. So, what of the awkwardness and discomfort on the part of myself and the German tourists? Emine's apparent unwillingness to connect with the tourists by making conversation more light-hearted than local politics appeared to alienate her guests and this made me feel uncomfortable. So perhaps my own discomfort was due to my sense that Emine did not understand what the tourists were seeking in their acceptance of her invitation, and also Emine's seeming inability, on her own at least, to provide a satisfactory experience for them. Indeed, the handicraft-selling part of the encounter was the most acutely uncomfortable, but also the most illuminating. At the time I was shocked that Emine did not approach her 'sale' with more subtlety and grace. I knew that with her husband physically displaying his lack of interest by not joining us on the floor to look at the items and instead remaining standing and gazing out of the window, the onus was on the German woman to take interest in and ultimately to buy the handicrafts. For my part, my being a woman of course helped because, as a woman, I was able to feign some interest in the overtly gendered offerings. However, I felt that the onus was thus on me to manage the sale and to make up for Emine's lack of grace.

Admittedly, my initial willingness to be involved in this encounter was primarily due to the general intent of my fieldwork being to explore the interactions between tourists and the local community and place of Göreme. In this particular period of fieldwork, moreover, a specific focus of my research was the apparent increase in recent years in the informal entrepreneurial activities of the Göreme women. I was keen to conduct participant observation in the arena of these activities wherever possible. It is inevitable, though, that the ethnographer who participates in serendipitous tourism encounters in this way finds her or himself within a complex set of reciprocal relationships. Indeed, I felt obliged to help Emine make the sale that she wanted from this meeting. I also felt compelled to help make the encounter satisfactory and enjoyable for the tourists. To do so, it was necessary for me somehow to prevent their feeling that they were being duped by Emine – and by me! It is in this pull of contradicting reciprocal relationships that my reflections upon the encounter expose the tensions in my uneven ethnographic knowledge and that I, as a 'tourism ethnographer', must reflect critically upon my own 'declarative' role (Hollinshead 2004).

TOWARDS RECOGNITION OF TOURIST SHAME

It is pertinent here to share a bit more detail concerning the process of the production of this article. By now I have taken versions of this paper to both an Anthropology of Tourism conference (Association of Social Anthropologists (ASA), London, 2007) and the 2nd Critical Approaches to Tourism conference in Split, Croatia, 2007. It was during the discussion of my paper at the ASA conference that a colleague and friend asked me why I had decided to side primarily with the tourists during the encounter. This question troubled me; I hadn't *decided to* do that, but that is the way it seemed to have turned out.

Although uncomfortable, this incident at the conference was utterly useful because it pointed me towards a re-interrogation of the discomfort I felt during the encounter and, consequently, towards a recognition of my own researcher, and tourist, shame. Conferences are, of course, a very embodied way for researchers to disseminate and create tourism knowledge and it was being there, face-to-face with my tourism researcher colleagues, that I was made shamefully aware of my double 'colonizing' positioning in the conduct of my ethnographic tourism research; in relation to Emine, I was both colonizing researcher (helping to make or define the world) and a colonising tourist (consuming and perhaps further appropriating a locality captured by and through tourism).

This shame-producing moment at the conference and my interrogation of this shame has thus led me to recognize that the discomfort during the actual encounter described at the start of this article was also caused by feelings of shame. Looking back, although I cannot quite remember fully, I recall that my torso became tense during the 'sale' part of the encounter and, at one point, I blushed. I also remember being quite unable to look the German woman straight in the eye at that time; I was acutely uncomfortable at the thought that the tourist couple might think I was helping Emine make a sale. My feeling of shame, moreover, was also in the broader recognition that I was not prepared to be more direct in helping Emine make a successful sale, and that I was being complicit and protecting of the othering 'tourist gaze'. By attempting to keep up some sort of pretence with the tourist couple that Emine was an authentic, uncorrupted cave-dwelling woman, I was not only using my neighbour and the hapless tourist couple to meet my own research ends, but I was engaging in worldmaking practices that were objectifying and essentializing each party in a way that locked them into an irresolvable problematic.

I can see now that, with our feelings of discomfort unrecognized and unacknowledged as shame, we, the Germans and I, continued throughout the remainder of the encounter to objectify Emine and to hold her within the frame of othering tourist discourse. I did so, for example, through my attempts to uphold her 'authenticity' regarding traditional food products in the food storage cave. The Germans did so by capturing in photographs the view from the terrace, the caves rooms in the house and by getting me to take a photograph of them with Emine against the backdrop of the house. There were also potential slippages in the upholding of the worldmaking gaze, with various opportunities to construct a more equal, embodied togetherness with Emine. Emine's tying the scarf on the German woman's head was a move that had the potential to equalize the pair in their gendered bodies. The shared moment of laughter regarding parental relationships was also a time of mutual understanding. However, these moments were not enough for the tourists' and my own discomfort to be in any way fully resolved. So, after we left Emine's house and walked away down the road, the German couple's comments about the encounter being 'very interesting' and a chance to 'get to see inside people's houses', served as a summarizing of the encounter with the 'other' and thus as a re-objectification of Emine. Our laughter about the German woman's wearing of a Göreme headscarf was perhaps to release some of the tension we were feeling regarding our objectification of the 'other', with her final gesture in taking the scarf off as we parted a futile attempt to remove this sense of discomfort so that she could continue on her way.

Although uncomfortable, the recognition of my own shame in this worldmaking/worldshaping incident has been utterly useful in that, as well as making me more aware of my own positioning in the encounter, it leads me to consider whether the discomfort that the tourists also clearly felt bore any relation to my own. Could it be that we all felt a sense of 'tourist shame' through sheer recognition of what we were doing in the encounter? As Probyn (2005: 78) argues, 'shame sets off a nearly involuntary re-evaluation of one's self and one's actions' (for a discussion of shame's relation to guilt, see Probyn 2005: 45–7). This re-evaluation of one's actions was shown clearly in the way that, when conceding to buy a headscarf from Emine, the German woman hung her head and said, 'I suppose they have to make money somehow'. For her, there might have been a double potential for shame: She might feel shame if she was not to buy something to help Emine make, at the very least after all, a meagre living. However, simultaneously, upon buying something she might take on the shame of a tourist being duped. Both of these elements of shame are uncomfortably antagonistic towards each other, but both also show how shame is brought on by a heightened awareness of the self as seen through the eyes of others. As Tomkins (1995: 136) suggests, shame 'is the most reflexive of affects in that the phenomenological distinction between subject and object is lost'. It is this distinction being lost that is the reason why shame 'should be considered crucially reflexive and embodied' (Waitt *et al.* 2007: 252) and this, in turn, opens up the postcolonial potentialities of tourist and tourism researcher shame.

POSTCOLONIAL POTENTIALITIES

If, as Spurr (1993) argues, travel narratives are discourses of colonialism through which one culture works to represent and to dominate another, then tourism inevitably works to define and fix both the tourist and the toured 'other' in a relationship that is always inherently colonial in nature. So, it is possible that the very 'politics of authenticity', i.e. the politics of nostalgia for a premodern world (Frow 1991) embodied in tourism encounters, like the one described in the present article, might evoke feelings of shame in any reflexive tourist who has an 'interest in living ethically' (Probyn 2004: 240). As Probyn (2005: 14) further points out 'Shame illuminates our intense attachment to the world, our desire to be connected with others, and the knowledge that, as merely human, we will sometimes fail in our attempts to maintain those connections'.

This is why shame, whilst uncomfortable, should be seen as positive in its reflexive and self-evaluative role. Indeed, shame in itself has an important worldmaking function in that it is a positive and productive highlighting of our interest and desire to live ethnically. In the tourism encounter shame thus functions as the 'specifically anti- or *post*-colonial *discursive* purchase in culture' that Slemon (1991: 3) defines as postcolonial.

Moreover, shame is both a bodily-based effect and, importantly, is shown in *bodily expression*: blushing, hanging of the head and inability to bring oneself to make eye contact. It is useful to draw further upon Probyn (2004; 2005) here and her argument that shame is 'simultaneously in our bodies, at the core of our selves, and in our social relations' (2005: 4). Of course, this is what renders debates about whether shame is an *emotion* or a bodily-based *effect* futile and, on this, Probyn cites shame as an example of Mauss' (1979: 10) discussion of 'physical and moral discharge'. Probyn reads Mauss's discussion of 'habitus' as going beyond Bourdieu's because of its analysis of 'the physiological, along with its place, its convergence with the psychological and the societal' (Probyn 2004: 235). Mauss manages, she argues, to convey the 'pervasiveness of physiological convergences' (Probyn 2004: 235):

> Mauss is careful to highlight the very thin layer that exists between the physiological and the social. Moreover that layer is inhabited and disturbed by the feeling body – its tears, laughter and ejaculations ... While Mauss also links the body's feelings to a 'precise end', he retains an emphasis on the very physiology which animates their moral discharge.

This 'very thin layer' between the physiological and the social is why feelings such as shame cannot be hidden and cannot be well controlled. Embodied tourism encounters can throw out surprises, not only to the other, but also to oneself; surprises that manifest in 'moral discharge' from the body. This in turn clarifies why it is that embodied tourism encounters cannot be reduced to issues of discourse alone, because encounters are always open to 'the potential to be disrupted by the fleshiness of the body' (Waitt *et al.* 2007: 252).

This bodily disruption to the flow of tourist discourse is directly where recognition of shame presents its postcolonial potentiality. Discussing a more overtly postcolonial context in that it is what McClintock (1992) refers to as a 'break-away settler society', Waitt *et al.* (2007) have looked at how tourist shame experienced by non-indigenous Australian tourists visiting Uluru (in central Australia) can open up a 'moral gateway' towards reconciliation with indigenous Australians. An example of this is in the following experience:

Through experiencing an Anangu (Aboriginal guide's) interpretation of country, Claire expresses a heightened respect for all indigenous Australians through an increased awareness of her own 'whiteness'. This consciousness arose through her body, from feeling shame about being white when participating in the Aboriginal guided tour. Rather than erasing shame through white racism of the frontier myth, the shame confronting the rewriting of colonization evoked for Claire the possibilities of co-existence in difference (Waitt *et al.* 2007: 259–60).

Thus, Waitt *et al.* (2007: 260) skilfully identify the way in which shame 'swings a moral gateway by providing ... a dispositional change, wherein the colonizing group alters behaviours to avoid repeating injustices that would be equally as damaging elsewhere'. However, for this 'moral gateway' to be fully opened it is necessary both for tourists to recognize and acknowledge their shame and for certain conditions, or skills, on behalf of the *toured* also to be present. Waitt *et al.* (2007) discuss the interpretative materials and guiding at Uluru that can provide these conditions.

In contrast, given the way that the encounter in Göreme described above unfolded, it is clear that the conditions were not present there for such a 'moral gateway' to open. Awkwardness and discomfort consequently dominated the experience as well as going on to dominate my memory (and perhaps the tourists' too?) of the experience. I can see now how I, in my positioning myself as go-between or 'culture broker' in an effort to ease my own discomfort and to try to make the interaction a success for all parties, achieved precisely the opposite. Indeed, as I said above, I managed quite spectacularly to lock the encounter into the irresolvable problematic of colonizing tourist discourse. Simultaneously, Emine, from her own position of shame and discomfort in being a Göreme woman in a tourism encounter, did not have the skill herself to unlock our felt – but at the time unrecognized – 'human moral concern for the other' (Cloke and Jones 2003: 200).

It might be asked here why the burden of dealing with and absolving tourist shame should be put on to the toured/locals/'hosts' at all. My answer to this is that it is only then, just as in the moral gateway opening up to reconciliation in Australia, that tourists and toured can begin to meet in a relationship of equality. Moreover, when hosts do know how to respond to and help the tourist work through their shame then they are in the position to choose whether they do so or not. If they do not possess such skill they are not in any position of control in the tourist–host encounter and the relationship remains one of worldmaking inequality. This becomes clear when considering how the encounters that tourists tend to have with Göreme men compare with encounters between tourists and Göreme women.

In my previous ethnographic work on Göreme, as well as discussing the gendered division of tourism there, I drew the general conclusion that the development of primarily small and micro- locally owned tourism businesses in

Göreme had led to direct economic benefits for the local community as well as a positive host–guest relationship between (male) villagers and tourists (Tucker 2003). In their tourism entrepreneurial activities, Göreme men have been interacting successfully with tourists and each other in the tourism spaces of Göreme for twenty years and are thus rich in know-how concerning tourists' imaginings of themselves. They are therefore not only highly competent in playing to those imaginings, but they are also able to play with them by engaging in ironic performances and caricatures of the tourists' images of themselves as cave-dwellers (see discussion of the Flintstones performances for tourists in Tucker (2002) and Tucker (2003: 169–80). Whilst the Göreme men would often tell me that they did not 'know' tourism in any formal sense, I always observed their high level of competence when it came to dealing with tourists, entertaining them and reflexively *playing to* tourist expectations whilst at the same time *emerging from* those expectations by somehow altering the power dynamics encoded in them. It is in this way that the Göreme men evoke Hannerz's (1990: 239) notion of cosmopolitanism, as marked by 'a personal ability to make one's way into other cultures, through listening, looking, intuiting and reflecting' as well as, in Vertovec and Cohen's (2002: 13) words, a 'built-up skill of manoeuvring through systems of meanings'. The men also evoke Hollinshead's (2004) relating of Bhabha's (1994) ideas on cultural hybridity to tourism, in that they present tourism's capacity 'to stand as a whole new medium through which subaltern peoples or emergent populations can experiment with the new lexicons of iconic identity as they creatively play at celebrating their felt "new", or even their felt "old", Selves' (Hollinshead 2004: 37). The Göreme men are indeed cosmopolitans or cultural hybrids in this sense since, through their high levels of competence and skill in their negotiations with their tourist guests, they manage to resist and repel the 'modernist understandings of being and identity which have hitherto tended to deprive them of their own subjectivities' (Hollinshead 2004: 36).

It has become clear in the present discussion that, beyond a cognitive understanding of tourist desire, it is necessary to also be able to read the 'moral discharge' from tourists' bodies, expressed in their movement and pose, in order to negotiate such successful tourism interactions. This in turn requires a certain reflexive understanding on the part of tourists' 'hosts', since it requires an understanding of one's own positioning and thus how to choreograph oneself opposite the tourist. Whilst the fact that this is a 'requirement' renders the requirement in itself colonizing, I would argue that the majority of the Göreme men who work in tourism do have such understanding and skill. Through their own repeated embodied interactions with tourists in Göreme's *pansiyons* (guest-houses), restaurants and other 'public' spaces, as well as through their ability to observe each other interacting with tourists and thereby share their tourism know-how, the men have learnt how to read tourists' moral discharge. They have learned how to recognize tourists' humility and shame, as well as how to recognize tourist shame as an 'interest in living ethically'. Indeed, those tourists who do not express this interest do not tend to be received graciously by the tourism entrepreneurs and do not stay long in the village (Tucker 2003). Those tourists who do admit humility and shame, in contrast, are rewarded with their hosts letting them know that it *is* possible to engage ethically in 'doing tourism'. This is done through the men's ironic and playful performances of hybrid subjectivity (Bhabha 1994; Hollinshead 1998a; 1998b; 2004), serving to work through the tourists' shame and thereby work the encounter free from its colonizing underpinnings.

The necessity of having repeated embodied encounters with tourists to develop such skill at understanding and reading tourist desire means that Göreme women, having only recently started to enter tourism spaces and engage in activities from which they have previously been excluded, lack the skill, or the 'soft things' (Puwar 2004: 110), to successfully negotiate these important 'declarative' or 'worldmaking' encounters. In Tucker (2007) I argued, borrowing Cone's (1995) term, that these entrepreneurial activities allow Göreme women to 'craft new selves'. In other words, they allow these women, to an extent, to develop 'emergent cross-genre identifications of Selfhood (and of Otherness)' (Hollinshead 2004: 37). I previously suggested that this process of 'crafting new selves' arises at the point that the women's 'hosting' tourists in their houses works to reconfigure 'domestic' space in its use for economic gain and that this goes hand-in-hand with a reforming of the way in which gender identities are performed, and thus 'made', through that space. However, it has become clear in the present discussion that it is precisely because the entrepreneurial endeavours of women like Emine remain relatively isolated in the confinement to 'domestic' space that women are far less able to develop a feel for the game, in Bourdieu's (1984) sense, of tourism. That is why the postcolonial potentialities of recognition and acknowledgement of shame remain, in many contexts, mere potentialities.

CONCLUSION

Shame is a curious emotion. In Tucker (2007), entitled 'Undoing Shame', I argued that it was necessary to undo shame in order for the Göreme women to be able to work comfortably in tourism space. That is because, in one of its formations, shame can be 'a powerful and potentially destructive and violent way to patrol the borders of normality' (Probyn 2005: xvi). In that sense, I viewed shame as a negative social process, which limited the ability of Göreme women to 'craft new selves' in the tourism arena. In this paper, however, I have argued for the positive, postcolonial potentialities of shame because, again in Probyn's (2005: 72) words, 'In shame, our habitus becomes reordered, shaken up, it admits other possibilities and in turn allows for more interest to be registered'.

We have seen the way that my own recognition and acknowledgement of shame has indeed shaken me up by prompting me to become aware of the unevenness of the ethnographic knowledge I create and also of the complicity in my ethnographic work with tourism's modernist representational practices. It thus becomes clear what Holmes and Marcus (2005:1101) are referring to in describing contemporary fieldwork as 'both slightly alienated and slightly paranoid in ways that are both inevitable and productive'. The inevitability of this 'paranoia' lies in the general increased reflexivity within critical Tourism Studies and the associated increased awareness among tourism researchers of their colonizing tendencies. This awareness is also highly productive in that it propels us to search for further possibilities to be critical of and in our research practices. For me here, my very bodily feelings of shame have led me to recognize my own 'declarative' functioning as a colonizing tourist-cum-researcher, this recognition itself being a worldmaking process.

Reflecting in this way upon my own discomfort in the encounter has thus highlighted the postcolonial potentialities of both tourist shame and tourism researcher shame. This identification of shame has been productive in that, first, it has illuminated my own and others' interest in living ethically. Furthermore, by looking closely at the emotional-affective dimensions of the encounter it has become clear how these dimensions inevitably disrupt the flow of tourist discourse and allow for new possibilities, including a more equitable relationship between tourists, toured and tourism researchers. However, in order for that potentiality to be followed through, it is necessary that all parties learn how to read the elements of emotional moral discharge from their own and each others' bodies. It is necessary, in other words, to develop an embodied competence in order to be able to reflexively play *to, with* and *against* colonial discourse in both worldmaking tourism practice and in worldmaking Tourism Studies. Following Bhabha's (1994) claim that the 'hybrid subject' offers the possibility of resistance to the totalizing repression of colonialism, recognition of emotion, and particularly shame, by tourism researchers, tourism 'hosts' and tourists themselves becomes a potentially important conduit for counter-narratives to the colonizing tendencies of tourism practice and Tourism Studies.

NOTE ON CONTRIBUTOR

Hazel Tucker (PhD in Social Anthropology at the University of Durham) is Senior Lecturer in Tourism at the University of Otago, New Zealand. She is the author of *Living with Tourism: Negotiating Identity in a Turkish Village* (Routledge, 2003) and co-editor of *Tourism and Postcolonialism* (Routledge, 2004).

REFERENCES

1. Andrews, H. 2006. "Consuming pleasures: Package tourists in Mallorca". In *Tourism, Consumption and Representation: Narratives of Place and Self*, Edited by: Meethan, K. and Miles, A. S. 217–235. Wallingford: CABI.

2. Ateljevic, I. and Hall, D. 2007. "The embodiment of the macho gaze in south-eastern Europe: Performing femininity and masculinity in Albania and Croatia". In *Tourism and Gender: Embodiment, sensuality and experience*, Edited by: Pritchard, A., Morgan, N., Ateljevic, I. and Harris, C. 138–157. Wallingford: CABI. [CrossRef]

3. Bhabha, H. 1994. *The Location of Culture*, London: Routledge.

4. Bourdieu, P. 1984. *Distinction*, London: Routledge.

5. Cloke, P. and Jones, O. 2003. Grounding ethical mindfulness for / in nature: Trees in their places. *Ethics, Place and Environment*, 6: 195–214. [Taylor & Francis Online]

6. Cone, C. A. 1995. Crafting selves: The lives of two Mayan women. *Annals of Tourism Research*, 22(2): 314–327. [CrossRef], [Web of Science ®], [CSA]

7. Crick, M. 1995. "The anthropologist as tourist: an identity in question". In *International Tourism – Identity and Change*, Edited by: Lanfant, M., Allcock, J. and Bruner, E. 205–223. London: Sage. [CrossRef]

8. Crouch, D. 2002. "Surrounded by place: Embodied encounters". In *Tourism: Between Place and Performance*, Edited by: Coleman, S. and Crang, M. 207–218. Oxford: Berghahn Books.

9. Crouch, D. 2005. "Flirting with space: Tourism geographies as sensuous / expressive practice". In *Seductions of Places: Geographical perspectives on globalization and touristed landscapes*, Edited by: Cartier, C. and Lew, A. 23–35. London: Routledge.

10. Frow, J. 1991. Tourism and the semiotics of nostalgia. *JSTOR*, 57: 123–151.

11. Hannerz, U. 1990. "Cosmopolitans and locals in world culture". In *Global Culture: Nationalism, Globalisation and Modernity*, Edited by: Featherstone, M. 237–251. London: Sage.

12. Hollinshead, K. 1998a. Tourism, hybridity and ambiguity: The relevance of Bhabha's 'third space' cultures. *Journal of Leisure Research*, 30(1): 121–156. [Web of Science ®], [CSA]

13. Hollinshead, K. 1998b. Tourism and the restless peoples: A dialectical inspection of Bhabha's halfway populations. *Tourism, Culture and Communication*, 1(1): 49–77.

14. Hollinshead, K. 2004. “Tourism and new sense: Worldmaking and the enunciative value of tourism”. In *Tourism and Postcolonialism: Contested discourses, identities and representations*, Edited by: Hall, C. M. and Tucker, H. 25–42. London: Routledge.

15. Holmes, D. R. and Marcus, G. E. 2005. “Refunctioning ethnography: The challenge of an anthropology of the contemporary”. In *Handbook of Qualitative Research*, , 3rd edn, Edited by: Denzin, N. and Lincoln, Y. 1099–1113. London: Sage.

16. Kandiyoti, D. 2002. “Introduction: Reading the fragments”. In *Fragments of Culture: The Everyday of Modern Turkey*, Edited by: Kandiyoti, D. and Saktanber, A. 1–21. New Jersey: Rutgers University Press.

17. Lengkeek, J. 2002. “A love affair with elsewhere: Love as a metaphor and paradigm for tourist longing”. In *The Tourist as a Metaphor of the Social World*, Edited by: Dann, G. M. S. 189–208. Wallingford: CABI Publishing. [CrossRef]

18. Massumi, B. 2002. *Parables for the Virtual: Movement, Affect, Sensation*, Durham: Duke University Press. [CrossRef]

19. Mauss, M. 1979. *Sociology and Psychology*, Edited by: Brewster, Ben. London: Routledge and Kegan Paul.

20. McClintock, A. 1992. The angel of progress: Pitfalls of the term ‘post-colonialism’. *Social Text*, 31/32: 84–98. [CrossRef]

21. Meethan, K. 2001. *Tourism in Global Society: Place, Culture, Consumption*, Basingstoke: Palgrave.

22. Pritchard, A. and Morgan, N. 2007. “De-centering tourism's intellectual universe, or traversing the dialogue between change and tradition”. In *The Critical Turn in Tourism Studies: Innovative Research Methodologies*, Edited by: Ateljevic, I., Pritchard, A. and Morgan, N. 11–28. Oxford: Elsevier. [CrossRef]

23. Probyn, E. 2004. “Shame in the habitus”. In *Feminism After Bourdieu*, Edited by: Adkins, L. and Skeggs, B. 224–248. Oxford: Blackwell Publishing.

24. Probyn, E. 2005. *Blush: faces of shame*, Minneapolis: University of Minnesota Press.

25. Puwar, N. 2004. *Space Invaders: Race, Gender and Bodies Out of Place*, Oxford: Berg.

26. Selwyn, T. 1996. "Atmospheric notes from the fields: Reflections on myth-collecting tours". In *The Tourist Image: Myths and Myth Making in Tourism*, Edited by: Selwyn, T. 147–161. New York: John Wiley and Sons Ltd.

27. Slemon, S. 1991. "Modernism's last post". In *Past the Last Post*, Edited by: Adams, I. and Tiffin, H. 1–11. Hemel Hempstead: Harvester Wheatsheaf.

28. Sönmez, S. 2001. "Tourism behind the veil of Islam: Women and development in the Middle East". In *Women as Producers and Consumers of Tourism in Developing Regions*, Edited by: Apostolopoulos, Y., Sönmez, S. and Timothy, D. 113–142. Westport, CT: Praeger Publishers.

29. Spurr, D. 1993. *The Rhetoric of Empire: Colonial Discourse in Journalism, Travel Writing, and Imperial Administration*, Durham, N. C.: Duke University Press.

30. Tomkins, S. 1995. *Exploring Affect: the selected writings of Silvan S. Tomkins*, Cambridge University Press: Cambridge. [CrossRef]

31. Tucker, H. 2002. "Welcome to Flintstones-Land: Contesting place and identity in Göreme, Central Turkey". In *Tourism: Between Place and Performance*, Edited by: Coleman, S. and Crang, M. 143–159. Oxford: Berghahn Books.

32. Tucker, H. 2003. *Living With Tourism: Negotiating Identities in a Turkish Village*, London: Routledge.

33. Tucker, H. 2007. Undoing shame: Tourism and women's work in Turkey. *Journal of Tourism and Cultural Change*, 5(2): 87–105. [Taylor & Francis Online]

34. Veijola, S. and Jokinen, E. 1994. The body in tourism. *Theory, Culture and Society*, 11: 125–151. [CrossRef], [Web of Science ®], [CSA]

35. Veijola, S. and Valtonen, A. 2007. "The body in tourism industry". In *Tourism and Gender: Embodiment, sensuality and experience*, Edited by: Pritchard, A., Morgan, N., Ateljevic, I. and Harris, C. 13–31. Oxford: CABI. [CrossRef]

36. Vertovec, S. and Cohen, R. 2002. "Introduction: Conceiving cosmopolitanism". In *Conceiving Cosmopolitanism: Theory, Context and Practice*, Edited by: Vertovec, S. and Cohen, R. 1–22. Oxford: Oxford University Press.

37. Waitt, G., Figueroa, R. and McGee, L. 2007. Fissures in the rock: Rethinking pride and shame in the moral terrains of Uluru. *Journal of the Royal Geographical Society*, 32: 248–263.

Chapter 3

The Role of Destination Attributes in Islamic Tourism

Mohamed Battour[1], Mohd Nazari Ismail[2]

[1,2] Faculty of Business and Accountancy, University of Malaya, 50600 Kuala Lumpur, Malaysia

[1] Faculty of Commerce, Tanta University, 31515 Tanta, Egypt

ABSTRACT

The objectives of this study are to test the relationship between tourism motivations and tourist satisfaction, and to test how 'Religion' moderates the relationship. The variable 'Religion' is represented by the availability of Islamic norms and practices which are relevant to tourism at the destination. The results of the Partial least square (PLS) indicated that the tourism motivations are significantly and positively related to tourist satisfaction. The results also showed that Religion significantly moderates the relationship between pull motivation and tourist satisfaction. However, the moderating effect of Religion on the relationship between push motivation and tourist satisfaction was not supported.

INTRODUCTION

Tourism industries are always searching for new segment of customers. For example, over the last decade the tourism industry has witnessed many firms in the industry catering to the needs of special groups such as elderly tourists, disabled tourists and gay tourists [1]. However, one relatively unexplored segment is the 'Religiously conscious' tourists. It is no wonder therefore that some researchers in this field insist that catering to the religious needs of any faith in this expanding industry is essential [2]. Although the relationship between tourism and religion has been addressed in the literature on tourism, there remains a shortage of theoretical publications in the area of destination attributes in the context of Islam. When it comes to the relationship between tourism and Islam, the lack of literature is more obvious, especially regarding Islamic norms and practices related to tourism at the destination and their impact on the needs of Muslim tourists. Thus, the objective of study is to test the relationship between tourism motivations and tourist satisfaction, and to test how 'Religion' moderates the relationship.

LITERATURE REVIEW

Religion is associated with tourism, in terms of consumer behaviour and the supplier, as well as the relationship between them [3]. However, there is limited research available on this relationship [4]. Destination marketing could generate greater satisfaction when appropriate tourists are targeted [5]. In response to that, the supply of religious attributes for Muslim tourists may lead to increased satisfaction.

This is an Open Access article distributed under the terms of the Creative Commons Attribution License 4.0, which permits unrestricted use, distribution, and reproduction in any medium, provided the original work is properly cited It is known that Muslim customers constitute a broad market which is around 1.82 billion around the world [6]. In addition, the Muslim market has its special requirements and its culture, which cannot be ignored. Thus, satisfying the religious needs of Muslim tourists may encourage them to travel to a specific destination. Religion is bound to be very important considerations when a Muslim decides to travel abroad [7]. The Muslim tourists may decide not to travel to a particular destination in the absence of some Islamic attributes [8, 9, 10]. This study attempts to address this problem by empirically testing the relationship between tourism motivation and overall tourist satisfaction with the availability of religion (Islam) as a moderating variable. The religion (INP) is represented by the availability of Islamic norms and practices related to tourism at the destination.

By using the travel motivation theory as a base, many researchers have tried to give more attention to the pull and push relationship by frequently modifying items associated with the constructs. However, very limited research is focused on empirically testing the relationship between tourism motivations and overall tourist satisfaction. To fill this gap, in addition to studying the tourism motivation for Muslim tourists, this research will also investigate the relationship between tourism motivations and the overall tourist satisfaction. The influence of travel motivations on overall tourist satisfaction has been studied in previous research [11,12]. Each variable, push motivations (PUSM) and pull motivations (PULM), have hypothesized effects on the overall tourist satisfaction (OTS). In addition, as a moderating variable, religion is proposed between tourism motivations and overall tourist satisfaction.

METHODOLOGY

A total of 1,300 questionnaires were distributed in international hotels and tourism sites in four Malaysian cities: Kuala Lumpur, Kuala Terengganu, Penang, and Johor Bahru. The questionnaires were handed out in these cities following a convenience sampling approach. There was a scanning question on the cover page of the questionnaire to determine whether or not the tourist was Muslim or not. If the tourist was a Muslim, he/she was invited to proceed in filling in the questionnaire. The partial least square (PLS) technique is selected to assess the structural model in the current study. In order to examine the relationship between the constructs in the proposed model as well as to test the hypotheses, two stages of analysis were performed to evaluate the structural model: (a) Structural model without the moderating variable, and (b) Structural model with the moderating variable. The effect size is also calculated, this being a measure of the strength of the theoretical relationship, including the moderating effect [13].

DATA ANALYSIS AND DISCUSSION

To develop INP as a construct, 18 items were generated based on the combination of conducting qualitative research and reviewing tourism and Islamic teachings literature. Out of 1,300 questionnaires, 551 were completed and returned from international hotels and tourism sites in four Malaysian cities: Kuala Lumpur, Kuala Terengganu, Penang, and Johor Bahru. After data cleaning, the final sample size was 508. The EFA results determined significantly the correlated factors, including six push travel motivations, five pull travel motivations, and four factors representing religion (See Table 3,4,5). After EFA, the measures were subjected to confirmatory factor analysis using AMOS 19 software. The results of the confirmatory factor analysis of the modified models of the push motivations, pull motivations, and religion are summarized in Table 1

Table 1. Fit Measures for the CFA Models

Fit Indexes	PUSM	PULM	INP	Acceptable level
Chi-square	203.582	95.886	304.933	
Degree of freedom (df)	94	48	129	
p	0.000	0.000	0.000	>0.05
Normed Chi-square (CMIN/DF)	2.166	1.998	2.364	< 3.00
Goodness-of-fit Index (GFI)	0.953	0.970	0.938	≥ 0.90
Adjusted Goodness-of-fit Index (AGFI)	0.932	0.951	0.918	≥ 0.90
Tucker-Lewis Index (TLI)	0.981	0.992	0.968	≥ 0.90
Comparative Fit Index (CFI)	0.985	0.994	0.973	≥ 0.90
Root Mean Square of Error of Estimation (RMESA)	0.048	0.044	0.052	≤0.08

A two-stage approach is followed for evaluating the interaction effects of moderating variables modelled in PLS(Chin et al., 2003; Henseler & Chin, 2010). This issue was investigated by comparing the baseline model in stage1 with the model in stage 2 (See Fig. 2). Results in stage 1(See Table 2) indicate that the direct paths, PUSM to OTS, PULM to OTS, and INP to OTS, are statistically significant as the t-values (3.85, 7.56, and 11.14) are greater than 1.64. The explained variance (R- squared) is 0.54 for OTS. The goodness-of-fit (GoF) is 0.56. Whereas the results in stage 2, compared with stage 1, show that the R-squared was increased to 0.66 providing evidence of a better explained variance. The effect size is 0.26 indicating that the Religion have more than a moderate effect (0.26 > 0.15) on OTS. Using procedures suggested by Tabachnick and Fidell (2007, p.148), the increased R- squared is attributable to the moderating effects and the effect size is statistically significant at 0.05 (F3, 504 =129.6 > Critical = 2.6). The GoF increased from 0.56 to 0.62, indicating a better fit of the Model in stage 2.

Table 2. Tests of PLS Path with Bootstrap for Stage 1 and Stage 2

	Stage 1		Stage 2	
	Path coefficient	*t*-value	Path coefficient	*t*-value
PUSM → OTS	0.11	3.85*	0.14	4.42*
PULM → OTS	0.33	7.56*	0.27	6.04*
INP → OTS	0.51	11.14*	0.49	10.34*
PUSM*INP → OTS	-	-	-0.11	1.25
PULM*INP → OTS	-	-	0.15	5.05*
R-squared	0.54		0.66	
GoF	0.56		0.62	
Effect size	-		0.26	

**Sig. if above 1.64 for 1-tailed test.*

Table 3. INP Factors

INP factors	Loading	Variance explained	Cronbach's alpha	Mean (Std. dv.)
Factor 1: Worship facilities		52.43	.883	3.76(.533)
Availability of Mosque (Masjid)	.752			
Availability of prayer facilities at tourism sites, airport, shopping malls, hotels, conference halls, parks, etc.	.726			
Presence of loud public pronouncement of Azan to indicate prayer time.	.761			
Placement of Qibla direction (Qibla stickers/direction point towards Makkah city) in your hotel room	.642			
Provision of a copy of the Holy Qur'an in each hotel room	.727			
Availability of water supply in toilets at tourism sites, airport, shopping mall, hotel, parks, etc.	.648			
Factor 2: Halalness		7.57	.893	3.76 (.53)
Availability of Halal food at tourism sites, airport, shopping malls, hotels, parks, etc.	.708			
Availability of segregated Halal kitchen in hotels and restaurants	.714			
Availability of segregated areas for women at beaches	.753			
Availability of hotels with segregated swimming pools and gymnasium for men and women	.697			
Banning of sex channels in hotel entertainment system	.747			
Factor 3: General Islamic morality		6.46	.889	3.78(.563)
Observation of Islamic dress code by hotel and restaurant staff	.624			
Prevalence of Islamic dress code (e.g. Hijab) at public places	.670			
Banning by the authority of prostitution	.779			
Banning by the authority of indecent display of affection between sexes at public places (such as kissing etc.)	.649			
Censorship by the authority of adult scenes in movies shown on TV	.745			
Factor 4: Alcoholic drinks and gambling free		5.70	.918	3.79(.606)
Banning of alcoholic drinks by the authority at public places (such as tourism sites, hotels, parks, etc.)	.857			
Banning of gambling activities by the authority at public places (such as tourism sites, hotels, parks, etc.)	.862			
Total variance explained		**72.18**		

KMO = .943 *Bartlett's Test of Sphericity* *p- Value =.000*

Table 4. Push Factors

Push factors	Loading	Variance explained	Cronbach alpha	Mean (Std. dv.)
Factor 1: Achievement		35.58	.950	3.21(.717)
Meeting new people	.808			
Going places friends have not been	.862			
Talking about the trip	.858			
Indulging in luxury	.829			
Factor 2: Exciting and adventure		7.42	.839	3.05(.589)
Finding thrills and excitement	.727			
Being entertained and having fun	.755			
Being daring and adventuresome.	.779			
Being free to act how I feel	.608			
Factor 3: Family togetherness		5.60	.942	3.24(.682)
Visiting places my family came from	.812			
Visiting friends and relatives	.774			
Being together as a family	.801			
Seeing as much as possible	.714			
Factor 4: Knowledge/education		5.43	.850	3.01(.712)
Learning new things or increasing knowledge	.721			
Experiencing new/different	.682			
Seeing and experiencing a foreign destination	.708			
Visiting historical places	.759			
Factor 5: Escape		4.828	.871	3.06(.677)
Getting away from the demands at home	.836			
Getting a change from a busy job	.828			
Feeling at home away from home	.664			
Experiencing a simpler lifestyle	.771			
Factor 6: Sports		3.728	.555	2.66(.710)
Participating in sports	.787			
Desire to watch sports events	.603			
Participate in physical activity	.758			
Total variance explained	**62.61**			

Table 5: Pull Factors

Pull factors	Loading	Variance explained	Cronbach's alpha	Mean (Std. dv.)
Factor 1: Natural scenery		35.71	.884	3.00(.806)
Outstanding scenery	.601			
Mountainous areas	.868			
Natural ecological sites	.867			
Wilderness and undisturbed nature	.839			
Factor 2: Wide space & activities		8.45	.875	2.95(.682)
Wide spaces to get away from crowds	.842			
Variety of activities to see	.604			
Water sports	.812			
Personal safety	.835			
Factor 3: Cleanness & shopping		7.74	.939	3.06(.699)
Standards of hygiene and cleanness	.754			
Shopping facilities	.810			
Reliance/privacy	.749			
High quality restaurants	.796			
Factor 4: Modern atmosphere		5.48	.784	2.92(.718)
Modern cities	.770			
Exotic atmosphere	.671			
First class hotels	.784			
Reliable weather	.749			
Factor 5: Different culture		5.06	.722	2.98(.611)
Interesting and friendly local people	.661			
Different culture from my own	.678			
Historic old cities	.734			
Interesting town/village	.750			
Total variance explained	**62.46**			

In stage 2, the moderating effect of INP on the relationship between PULM and OTS is statistically significant as the t-value of the interaction path is 5.05 and higher than 1.64. However, the results show no support for the moderating effects of INP on the relationship between PUSM and OTS as the t-value of this path is 1.25 and less than 1.64. These results suggest that the Model in stage 2 is better than the Model in stage1 and provides evidence that the Religion moderate the relationship between pull motivation and the overall tourist satisfaction. Figure 1 shows that the results supported hypotheses H1, H2, and H4, as evidenced by the path coefficient and significant t-value. While hypothesis H3 was not supported meaning that there is no significant moderating relationship between PUSM and OTS by INP.

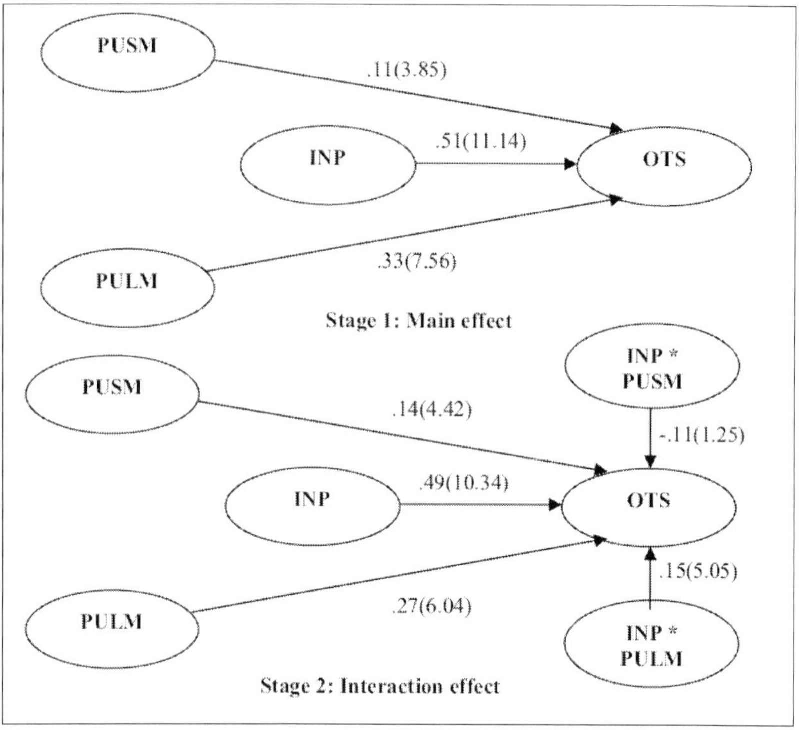

Figure 1. Testing Moderating Variable

CONCLUSION

The results show that four factors were extracted from the eighteen Islamic norms and practices items and labelled; worship facilities, Halalness, general Islamic morality, alcoholic drinks and gambling free. Lack of public consumption of alcohol and public gambling activities were found to be the most important Islamic norms and practices with a mean score of 3.79, followed by general Islamic morality (3.78), worship facilities (3.765), and Halalness (3.763). The results of the PLS indicated that religion significantly moderates the relationship between the pull motivation and the overall tourist satisfaction based on the significant interaction and variance explained (= 0.15, t-value (5.05) >1.64). This result confirms that the availability of Islamic norms and practices related to tourism at the destination contributes to the overall tourist satisfaction and strengthens the relationship between conventional destination attributes and tourist satisfaction for Muslim tourists. By having Islamic norms and practices, destination marketers can achieve tourist satisfaction among Muslim tourists. Furthermore, if destination marketers determine the preferable Islamic norms and practices to Muslim tourists, this could help them to design and tailor Halal products and services. The findings of this study should help marketers to better understand "Islamic tourism" and to develop marketing strategies to attract Muslim tourists. Furthermore, destination marketers might be able to design creative programmes that harness the unique characteristics of tourism products to satisfy and delight Muslim tourists. For example, design resorts fully meet Islamic religious needs such as segregated beaches, segregated swimming pool, and alcoholic drinking free.

REFERENCES

1. Weidenfeld. Religious needs in the hospitality industry. Tourism and Hospitality Research,**6**,2, 143-159 (2006)
2. Weidenfeld & A. Ron. Religious needs in the tourism industry. Anatolia:international journal of tourism and hospitality research, **19**,2, 18-22 (2008)
3. Y. Poria, R. Butler & D. Airey. Tourism, Religion and Religiosity: A Holy Mess. Current Issues in Tourism, **6**,4, 340-363 (2003)
4. A.C. Howe. Queer pilgrimage: The San Francisco homeland and identity tourism. Cultural Anthropology, **16**,1, 35-61 (2001)
5. M. Wheeler. Tourism marketing ethics: an introduction. International Marketing Review, **12**,4, 38-49 (1995)
6. Muslim population worldwide. Retrieved 15 November, 2013, from http://www.islamicpopulation.com/ (20130
7. WTM. The World Travel Market Global Trend Reports 2007. World Travel Market:London (2007)
8. M.M. Battour, M.N. Ismail & M. Battor. Toward a Halal Tourism Market. Tourism Analysis, **15**,4, 461-470 (2010)

9. M.M. Battour, M.N. Ismail & M. Battor. The impact of destination attributes on Muslim tourist's choice. International JM. ournal of Tourism Research, **13**,6, 527-540 (2011)

10. M. Battour, M. Battor and M.A. Bhatti. Islamic Attributes of Destination: Construct Development and Measurement Validation, and Their Impact on Tourist Satisfaction. Int. J. Tourism Res.. doi: 10.1002/jtr.19479 (2013)

11. C.G.Q. Chi & H. Qu. Examining the structural relationships of destination image, tourist satisfaction and destination loyalty: An integrated approach. Tourism Management, **29**,4, 624-636 (2008)

12. M. Devesa, M. Laguna & A. Palacios. The role of motivation in visitor satisfaction: Empirical evidence in rural tourism. Tourism Management, **31**,4, 547-552 (2010)

13. W.W. Chin, B.L. Marcolin & P.R. Newsted. A partial least squares latent variable modeling approach for measuring interaction effects: Results from a Monte Carlo simulation study and an electronic-mail emotion/adoption study. Information Systems Research, **14**,2, 189-217 (2003

Chapter 4

The Untapped Potential of Sustainable Domestic Tourism in Iran

Habib Allpour*, Hasan Kilic[1] and Noushin Zamani[2]

Faculty of Tourism, Eastern Mediterranean University, TRNC Via Mersin 10, Gazimagusa 95, Turkey

ABSTRACT

The majority of tourism literature focuses on international tourism; however, domestic tourism for specific destinations, such as Iran, can contribute to job creation, heritage protection, regional integration, and cross-cultural understanding. The market for domestic tourism in Iran has been downplayed and its true value has been underutilized. This study assumes that the present apathy towards tourism is reinforced by the public sector's lack of decisive governance towards the tourism industry where the public sector "can best act to mediate contemporary tourism-related social, economic, political and environmental policy problems" (Hall, 2011, p. 439). Drawing on "culture-ideology of consumerism" and "ecological economics" theories, study revealed that critiques of international mass tourism have convincing argument that domestic tourism is a domestic investment with a sustainable agenda.

Keywords: domestic tourism, tourism planning, global tourism, sustainable tourism

INTRODUCTION

Iran's tourism potential makes it a highly suitable candidate for renewing the focus on tourism resources in order to enable it to reap its share of the world tourism "pie". However, the industry has been neglected for various political and social reasons, especially since the Islamic revolution of 1979. As a result, the tourism industry has remained a low key activity in comparison to other sectors. In the wake of the post-revolutionary upheaval, Iran experienced a rapid population growth due to the adoption of a (pronatalist) policy and the suspension of family planning, which has resulted in a huge rise in the fertility and population growth rates during the first decade after the revolution (1979–1989) (Mehryar & Ahmad-Nia, 2004). Demographic shifts have been accompanied by higher transportation use – mainly car ownership – which has allowed for increased mobility of the population. These developments have led to an increase in both domestic and outbound tourism. Iran, with a large population, in the younger age groups of 15–29, is facing controversial issues in leisure and tourism (Sheykhi, 2004).

In the meantime, Iran's economy is highly dependent on the export of fossil fuels; with unemployment running in double digits, the tourism industry has remained dormant, in spite of its tremendous potential for a variety of tourism resources. With the future of non-renewable energy sources in doubt, tourism remains a prospective alternative to combat economic and social hardships. In spite of the lack of formal and institutional attention to the tourism industry, it remains that both domestic and visiting friends and relatives tourism have been established in Iran for some time (O'Gorman, Baum, & McLellan, 2007). To date, domestic tourism in Iran has revolved around pilgrimages to the shrines of revered Shia saints and holidays to the Caspian Sea. As a result, the spatial pattern of tourism in Iran is characterized by its concentration at the few pilgrimage sites and the Caspian Sea region. Demand of the latter is high as represented by leisure seekers from mainly Tehran and the rest of the country.

The study aims to fill some of the existing gaps in the literature in regard to domestic tourism and its implications for a sustainable tourism specific to certain destinations. This is tantamount to local tourism governance, which is a characteristic of domestic tourism – external to global tourism system (Higgins-Desbioless, 2012; Sklair, 2010).

THE SUSTAINABILITY DISCOURSE

The overarching concept of this study is the "sustainability" discourse and its significance for planning (Bianchi, 2004; Hall, 2011). As planning has evolved to respond to the changes in human communities, sustainability has become an instrumental concept to frame the planning process when the basics of societal or environmental welfare are threatened (Hanna, 2005). It is also a reactive idea to respond to the market failures in the context of global and local arenas. As Hall (2011, p. 447) states:

nevertheless, the failure to achieve desirable outcomes as a result of self-regulation, market failure and the limits of the market as a form of governance has increasingly been recognized, especially with respect to the equity of policy outcomes and the achievement of more sustainable forms of tourism.

At any rate, sustainability discourse has come of age. As Saarinen commented, "sustainability has become a discourse which challenges and directs the political and economic structures that constitute the larger context of the tourism industry and tourism development" (as cited in Bianchi, 2004, p. 495).

Sustainability as a concept has appeared as a fundamental principle of changing dimensions of tourism planning (Hall, 2008) that evolved from "boosterism" to *community-based* tourism and eventually to the presently hotly debated "sustainability" discourse (Dodds, 2012; Hall, 2008). However, sustainability has not remained a theoretical discourse; rather, it has become a form of governance, especially in tourism planning (Hall, 2011). As Hall (2011, p. 451) comments, "governance will clearly be a significant concept in tourism destination planning and policy in the foreseeable future".

Besides the concept's popular/general definition that warns us not to compromise the quality of the resources for the sake of future generations (World Commission on Environment and Development [WCED], 1987), *sustainability* has evolved into a paradigm (Martinez-Alier, Pascual, Viven, & Zaccai, 2010) encompassing issues that

interwoven in a complex debate on how best to achieve equity of access to natural resources which create human well-being, and in the distribution of the costs and benefits (social, economic, and environmental) which ensue from the utilization of resources. (as cited in Hunter, 1997, p. 851)

The assumption is that mass tourism is not necessarily sustainable – at least not in every destination – because it does not operate according to the principles of sustainability laid down in the report "our common future" (WCED, 1987) and *Agenda 21*. The concept was theorized in "our common future" and articulated further towards practicality in 1992 at the United Nations Conference on Environment and Development in Rio (the "Earth Summit").

As tourism has been divided into two fields/constructs (i.e. "impact" and "business"), the sustainability paradigm has become instrumental in and a device for understanding the anatomy of the "tourism impact" on all aspects of destinations around the world (Burns, 1999; Tribe & Airey, 2007). In fact, it is the conflict between business interest and impacts (i.e. environmental impact, tourism perception, carrying capacity, and social impact) that ensued sustainable tourism definition as "tourism which is economically viable but does not destroy the resources on which the future of tourism will depend, notably the physical environment and the social fabric of the host community" (as cited in Saarinen, 2006, p. 1124). And by the 1990s, many analysts acknowledged the power of tourism as a social force (Mura & Tavakoli, 2012) and an emerging paradigm of sustainability with its principles as the cornerstones for the maintenance and growth of tourist destinations.

The nexus between global capitalism and international mass tourism legitimated concerns and raised questions whether tourism can remain immune to the global consumption pattern, which is based on profit maximization and is contradictory to ecological carrying capacity (i.e. critical limits) through exploitation of natural resources in the periphery (Dauvergne, 1997). Adherents to the sustainability paradigm emphasize "that there are concrete limits to ecological capacity and that given ecosystems have finite carrying capacities; therefore there is a need to preserve these ecosystems and respect their limits or face economic and environmental collapse" (as cited in Carvalho, 2001, p. 63). They are specifically critical in regard to the nexus between international mass tourism and global capitalism as they consider the extraordinary capacity and power of tourism to be instrumental in intensification of "culture-ideology of consumerism", which undermines sustainability (Higgins-Desbiolles, 2010; Sklair, 2010). To put it differently, the present outlook of global capitalism and its production/consumption agenda has also changed. Hardt and Negri (2004, p. xii) articulated by stating:

> ...the recognition that contemporary global order can no longer be understood adequately in terms of imperialism as it was practiced by the modern powers, based primarily on the sovereignty of the nation-state extended over foreign territory. Instead, a "network power", a new form of sovereignty, is now emerging, and it includes as its primary elements [*international mass tourism*], or nodes, the dominant nation-state along with supranational institutions [*WTO, WB, IMF, ILO, TNPs*], major capitalist corporations [*tourism transnational corporations*], and other powers.

It achieves this by focusing on four significant indicators/dimensions to explain market failures pertinent to the conventional economic growth perspective (Hosseini, 2003). First, sustainability is a holistic approach that moves beyond conventional economics by constructing "ecological economics" (Loomis, 2000). The emergence of the sub-discipline of "ecological economics", which claims to address the shortcomings of conventional economic analysis, is a step in this direction (Driml & Common, 1996; Loomis, 2000). Second, it does this by valuing natural capital (i.e. natural resource input), which includes natural stocks in the context of the concept of a capital asset that was ignored by conventional economics. Third, it perceives social/cultural capital as a significant element of development. Fourth, it invests in a "balanced portfolio of stocks by having established the constant rule of capital, sustainability conditions for production and consumption, and a broader set of stocks or resources than mainstream economic" (Loomis, 2000, p. 895).

MASS TOURISM AS A GLOBAL PROJECT

The question is whether a global tourism structure is organized and operates according to a "development-first" perspective – which is holistic and sustainable – or to a "tourism-first" paradigm – which is market dominated/driven (Burns, 1999). The two approaches to tourism planning, pertinent to the so-called Third World countries, may be placed on a continuum – as polar opposites to each other (Burns, 1999). Depending on which approach is the preferred case, the industry itself becomes the subject of scrutiny in the context of a global tourism structure with respect to sustainability, which is at the heart of the discourse (Higgins-Desbioless, 2009; Wheeler, 2012). Thus, the concept and aim of development via "tourism" takes a new twist within the globalization discussion. Sklair (2010) eloquently poses a thorny dimension hitherto ignored by the voluminous tourism literature that is focused on micro-level studies that are highly charged with neo-liberal economic perspectives (Higgins-Desbiolles, 2009; Martinez-Alier et al., 2010). Sklair's view, which is disturbing to neo-liberalists' ears, is at the heart of this discussion. He touches upon the most visible and tangible problem associated with global capitalism, as well as globalized tourism, in the following terms (Sklair, 2010, p. 115):

Capitalist globalization simultaneously enriches more people all over the world than ever before in human history, but cannot bring the poorest out of debilitating poverty in most societies, and those in the middle seem fated to suffer cyclical bouts of economic insecurity, and the crisis of ecological unsustainability – capitalist globalization spreads industrial civilization all over a planet that cannot cope.

No doubt, there are certain positive impacts of the globalized tourism phenomenon; however, the negative social externalities of mass tourism to the domestic residents' welfare cannot be ignored. One of the indications of social externalities is the "consumption congestion" (Chao, Hazari, & Sgro, 2004), which eventually lowers the positive impacts as the number of tourists increases. Chao et al. (2004) utilized a *general equilibrium* model to analyse the negative social externalities in relation to non-tradable goods in the domestic economy of the destination with respect to competition. Imperfect competition and negative social externalities are also highly relevant to disorganized capitalism, which examines the economies of space and produces many more cultural artefacts or signs than people can cope with (Lash & Urry, 1994). However, this happens through the mobilization of objects (i.e. money, productive capital, and commodities) and subjects (i.e. capital and labour power). But what corresponds to disorganized capitalism is a "tourism" industry that becomes an instrument of "fixes" (i.e. intrinsic to the development of the international tourism industry) (Fletcher, 2011) to global capitalism. Therefore, based on Urry and Lash's analysis (1994, p. 17), tourism becomes a salvation to the spatial institutions of capitalism as articulated in following terms:

> The old order [*i.e., Fordist*] has been significantly undermined by two processes. The first is the disintegration of the old core with finance, distribution, property, service, and knowledge and R&D functions each taking on their own autonomy. The second is the formation of a new core, one in which "the post-industrial tail of the old order begins effectively to wag the Fordist and industrial dog". The new core is clustered around information, communications and the advanced producer services, as well as other services, such as telecommunications, airlines and important parts of *tourism and leisure*.

The role of tourism, among many other forms of communication (media, Internet, TV stations, satellites, etc.), as a facilitator of transnational interactions' involvement becomes clear when "...the cosmopolitan–local continuum takes the form of different degrees of attachment to specific locales, countries, local cultures and communities, and finally, to the national economy" (Roudometof, 2005, p. 125). Not to mention the corporate lobbies that have campaigned to water down numerous sustainability initiatives aimed to curb the activities of multinational corporations. And/or, "the World Travel and Tourism Council (WTTC) [*that*]has consistently campaigned against aviation fuel taxes and all forms of government regulation which might negatively affect the balance sheets of its members" (as cited in Bianchi, 2004, p. 498).

The issue of international tourism consequently becomes a global project whereby global capitalism enjoys its utility for the purpose of expansion of capital by means of internal globalization (Beck, 2002) or glocalization. In this regard, Urry and MacCannell commented that

> paradoxically, while many subjects look for authentic sites of real experience, the industry clones cultures, cities, and feelings to connect tourists with nothing (emptied spaces). Based on the assumptions that culture and tourism facilitated the expansion of capital, sociology and anthropology should reveal the relations of power enrooted in mobility. (as cited in Roudometof, 2005, p. 3)

By focusing on transportation and its global connotation (Lash & Urry, 1994), the nexus between mass tourism, environment, and energy crystallize into a global political economy. As Mitchell (2002, p. 259) stated

> the demand for oil for transport is critical to the long-term growth of oil. Transport accounts for nearly half of oil demand today and by 2020 may account for around 55%. The projected increase in transport demand accounts for three quarters of the conventional vision projections for a 50% increase in oil demand by 2020. (Mitchell, 2002, p. 259)

DOMESTIC TOURISM: A SUSTAINABLE OPTION?

With respect to the significance of tourism in the global economy and that of individual destinations, the concept of tourism addresses two major sectors: "international" and "domestic". Domestic tourism has been defined as "...a form of tourism, involving residents of the given country travelling only within this country" (World Tourism Organization [WTO], 1995, p. 11). Notwithstanding the paucity of studies and research on domestic tourism dynamism worldwide, the UK residents alone took 76.8 million domestic holidays within the UK in 2007, spending £14 billion. England accounted for 60.9 million of these holidays and £10.9 billion of spending. In Wales, the equivalent figures were 6.5 million holidays and £1.1 billion, and in Scotland, 8.7 million holidays generated £1.9 billion. Domestic tourism for the UK registered 126.0 million trips (an increase by 7%) for the year 2009 (http://www.britishresorts.co.uk/tourismfacts.aspx). In 2005, domestic tourism contributed an estimated $55.5 billion to the Australian economy, more than three times the contribution of international arrivals. Domestic tourism has also made various impacts in many areas of China. "It makes a significant contribution in promoting regional economic growth, improving local economic structures, driving the development of related industries, enhancing employment and activating domestic demands" (Wu, Zhu, & Xu, 2000, p. 296). Domestic tourism in Turkey has experienced burgeoning growth between the 1980s and 2000 such that the number of internal travellers has reached 16.4 million. This growth has been attributed to an increasing income level, improved modes of transport and transportation facilities, cultural change, and an improved accommodation sector (Demir, 2004, p. 327).

The WTO predicts that during the next 20 years, the expansion of domestic tourism will be especially strong in several developing countries, most notably China, India, Thailand, Brazil, and Mexico. The scale of world domestic tourism, on the other hand, exceeds the world international tourism by a ratio of 10:1. Despite domestic tourism's role in the socioeconomic and infrastructural development of its various destinations, it has not received sufficient attention in terms of research and analysis. In the meantime, through the lenses of "critical inquiry", domestic tourism fits into the sustainability paradigm argument in contrast to international mass tourism, whereby the latter is highly criticized for being an instrument of operationalization of "culture-ideology of consumerism" (Sklair, 2010, 2002), which stands as a paradox to sustainability. With respect to sustainability, domestic tourism also fulfils another objective of the discourse (i.e. justice tourism) (Higgins-Desbiolles, 2008) in the form of "social tourism"/tourism for all, which is more feasible in the case of domestic tourism. As Higgins-Desbiolles (2010, p. 126) articulated,

for instance, holidaying closer to home may provide us with greater connections to our community and its environment, thereby combating the sense of alienation increasingly evident in our globalizing world and fostering an appreciation of our own environment that is conducive to a commitment to sustainability.

THE ADAPTIVE DOMESTIC TOURISM MODEL

An adaptive domestic tourism model (ADTM) is illustrated in Figure 1, which contextualizes domestic tourism planning within a sustainability framework that accounts for the resources, activities, and communities to be integrated in each domestic tourism enterprise locally and regionally (Larson & Poudyal, 2012). An ADTM is adopted to fit Iran's case with respect to culture, religion, geography, demography, environment, and the Iranian political economy. The model focuses on domestic tourism development in order to remove ambiguity regarding the goals and strategies that must be clarified by policy-makers and planners. Therefore, an ADTM is a framework that can facilitate the acquisition of knowledge in this process. It is also essential to clarify the potential and problems at the beginning of the process through the feedback loop. Notwithstanding national decisions at the macro-level, a National Tourism Organization (NTO) is a prerequisite to establish a functional institution dedicated to tourism (i.e. contrary to the present NTO that is not built to focus on domestic tourism). This will clarify and facilitate a political vision "that step[s] beyond a sometimes fragmented and disorganized administrative culture" (Burns, 2004, p. 33). An NTO for domestic tourism will have three fundamental objectives: first, it will remedy some portion of the economic ills through job creation; second, it will provide an uplifting recreational means for a population in desperate need of a transition from revolution to recreation and regeneration; third, it will facilitate sociocultural harmony through interaction among ethnicities that can bring cohesion and understanding across different ethnic groups in a multi-ethnic country (Moufakkir, 2010).

PowerPoint slideOriginal jpg (594.00KB)Display full size

In a way, the "resource-based" tripod lays the foundation for three levels (indicated in grey bubbles) at which domestic tourism will function. In general, characteristics of each level are different in terms of capacity, education, attractions, and the host. Such scaling will allow for better control, management action, implementation, and monitoring. This is also conducive in Iran's case, which is challenging geographically and culturally. It is tantamount to a decentralized approach that will give innovative ideas based on knowledge of the community at each level (Brouder, 2012). In addition, the ADTM is not a top-down planning process, but rather a place-based model guided by local knowledge and concerns (Larson & Poudyal, 2012).

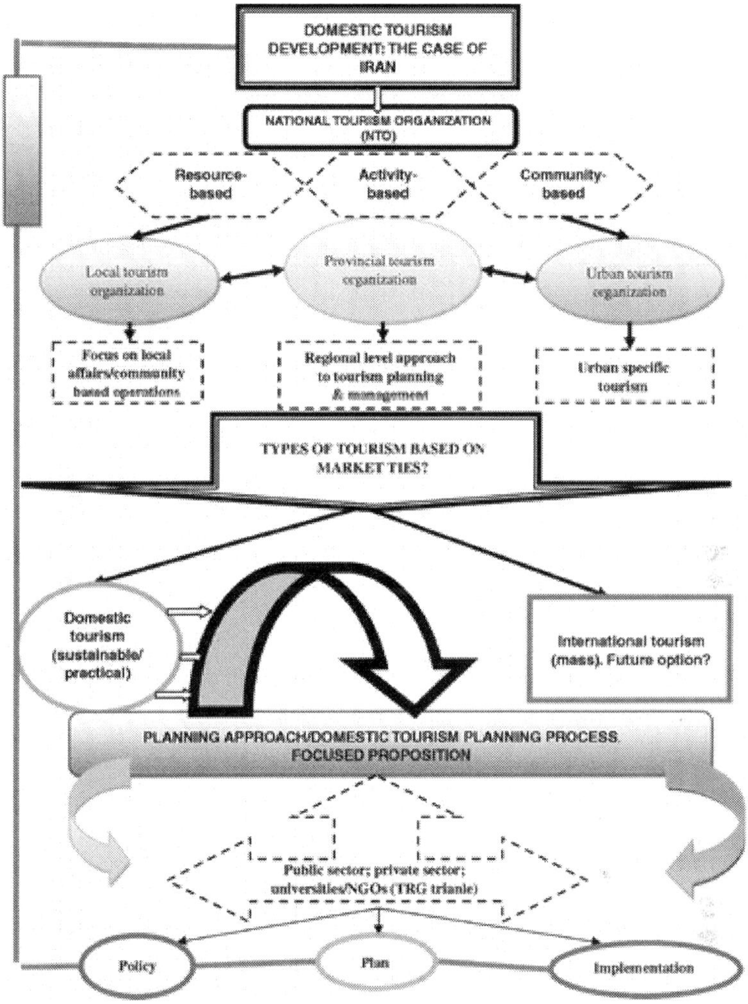

Figure 1 An adaptive domestic tourism model.

ADTM focuses on market ties, which, in the case of domestic tourism, takes on a different meaning and morphology. In the case of Iran, which is a highly multicultural and multi-ethnic entity, a rich market with a variety of motivations to travel at home also exists. ADTM recognizes the complexity of a tourism system that involves many aspects of society and numerous scientific disciplines. Thus, it calls for an embodiment of a symbiotic triangle (i.e. tourism, research, and governance) in the context of NTO.

However, the decentralized nature of the model aims to facilitate the flourishing of civil institutions, specifically for tourism that is a new introduction to traditional social structure and an economy based on primary resources. This aspect is a deficit in the case of Iran, as well as in most developing countries (Burns, 2004).

An ADTM is not a "one-shot, big bang" (Burns, 2004) master plan that is highly technical and discrete; rather, it is based on local values with measurable outcomes and immediate, mid-term, and long-term utility. The philosophy underpinning an ADTM is for it to be implementable in a way that takes full account of the local social dynamics in the context of a platform for sustainable growth and human development. Nevertheless, the public sector's role in the specific case of Iran remains paramount to implement ADTM, because "large-scale voluntary improvements in sustainability are improbable", especially given how the private sector behaves at the start of tourism development and the "particular ambivalence to tourism" (Buckley, 2012, p. 535).

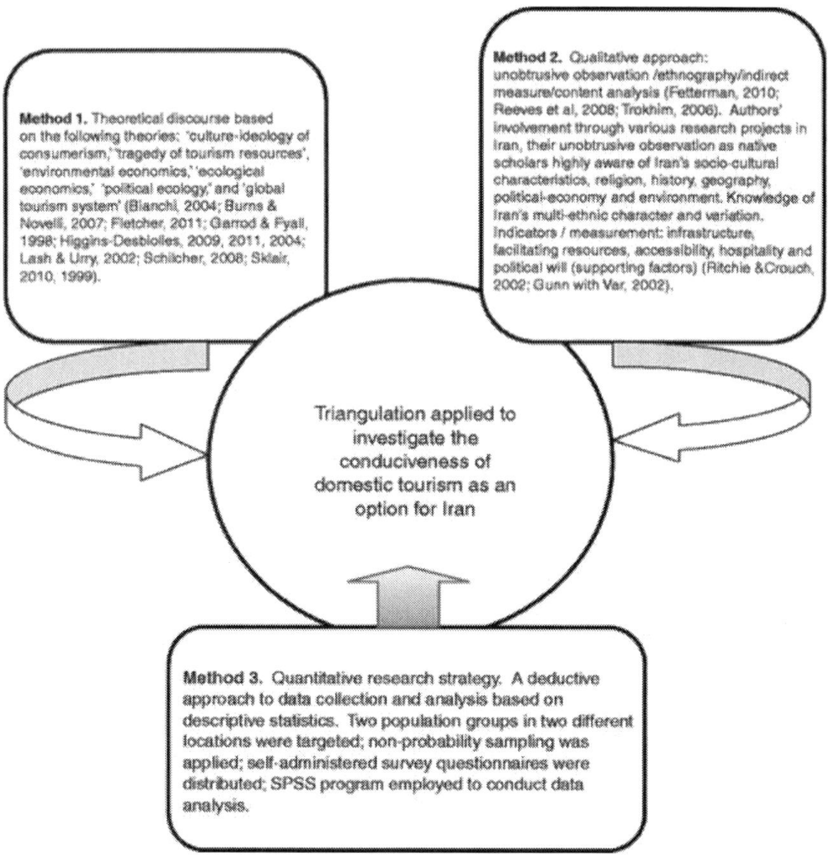

Figure 2. A triangulated research process model.

METHODOLOGY

The study has taken a multi-strategy approach known as triangulation, which entails using more than one method of data collection in the study of social phenomena (Fetterman, 2010) (see also Figure 2). The first research prong (Method 1) brings in sociological theories of the "culture-ideology of consumerism", the "tragedy of tourism resources", "environmental economics", "ecological economics", "political ecology", and the "global tourism system" (Fletcher, 2011; Higgins-Desbiolles & Whyte, 2013; Lash & Urry, 1994; Schilcher, 2007; Sklair, 1999, 2010).

PowerPoint slideOriginal jpg (396.00KB)Display full size

Each of these theories brings sociological, ecological, and political-economy discourse into the context of global capitalism and the global economic system that critically analyses the paradoxes between the international tourism system (mass tourism) and sustainable development. The above authors argue that sustainable tourism development is a far-fetched goal because the tourism industry is interwoven with global capitalism, based on overproduction and overconsumption, such that environmental and social problems are an inevitable result.

In the specific case of Iran, domestic tourism can be an acceptable alternative to fulfil the demand of travellers and will also remain immune to the sanctions inflicted by global capitalism. For instance, the sanction on selling passenger airplanes to Iran has already inflicted a devastating impact upon Iran's tourism industry. This situation has been exacerbated by official's perception of international mass tourism as a global conspiracy to export western decadence to the Islamic Republic of Iran. "... [in fact] a significant part of the influence of Western culture on Islamic societies comes through leisure time activities: watching television, using the internet, playing and watching sport, hearing Western music, are obvious examples of this cultural impact. Such trends usually lead to socio-cultural controversies in Iran" (cf. Sheykhi, 2004, p. 63). Therefore, an alternative form of tourism (e.g. domestic tourism) may be sustainable and averse to the hazards of globalized capitalism (Burns, 2004; Higgins-Desbiolles & Whyte, 2013). This is also defended in respect of the present-world economic crisis and its implications for international mass tourism (Papatheodorou, Rosselló, & Xiao, 2010).

The second research prong (Method 2) applied an inductive approach based on observation, content analysis, face-to-face interviews, researcher reflexivity, and numerous visits and personal dialogues with officials in Iran. Within the second prong, face-to-face interviews contributed to the analysis of field data (Trochim, 2006). Face-to-face interviews were conducted with officials in Iran Cultural Heritage, Handcrafts and Tourism Organization (ICHTO). ICHTO is the

Iran's NTO overseeing numerous other governmental and quasi-governmental affiliates ranging from Iran Air (Iran's national carrier) to several heritage and culturally related organizations. In all, 17 officials were interviewed based on an open-ended approach. Each item was based on thorough reviews of the literature on sustainable development, sustainable tourism, international tourism, domestic tourism, tourism planning, tourism development and planning in Iran, as well as the political economy of sustainable tourism.

The interviews were conducted in conjunction with informal conversations with travel agency managers and hoteliers as well as academicians. All interviews were recorded in the field with a voice recorder. They were later transcribed from the recordings, then categorized, and coded. Nvivo 10 was utilized to perform data analysis and refine the data, which is based on five stages: source, coding, nodes, source classifications, and node classifications. The analysis revealed that officials were familiar with the concept of sustainability; however, their perceptions were devoid of knowledge of its operationalization. Overall, there is a lack of consensus among the policy-makers regarding a formidable commitment to full force tourism development because of ideological dispute among the layers in the formal sector. For instance, collaboration in relation to tourism planning and development in Iran is not necessarily a process of joint decision-making among key stakeholders.

Governance for tourism, as the study revealed, is a mixture of approaches that does not correspond to what Hall (2011) articulated under a typology of tourism public policy. At best, a hierarchical type (Hall, 2011) is relevant to Iran's case, which is not necessarily compatible to a "sustainability" paradigm (Larson & Poudyal, 2012). Study revealed that in the specific case of Iran, which can only be categorized as a hierarchical type of governance (Hall, 2011), distortions occur as the policies do not flow harmoniously. This means that when a decision is made at a level in the hierarchy (top-down steering), it will not necessarily be implemented. This is because, in the structure of the hierarchy, there is always one layer (i.e. due to multiplicity of the power base) that may not agree with the policy and its implementation. For instance, if a municipality is deciding to improve the attractivity of its city by application of street decoration or pedestrian facility development, it can be easily blocked by a religious authority as a decadent project (Mayor of Tabriz, personal communication, May 2013). This is also in line with Hall's (2011) assertion that there is a better chance of implementation of public policies in a non-hierarchical type of governance.

The third research prong (Method 3) focused on domestic travellers' perceptions in order to explore their views about the difficulties associated with travelling in Iran. The assumption is that travellers' perceptions can shed some light on the behaviour of the public sector towards domestic tourism development. For this purpose, two locations were targeted based on their ranking in terms of the number of domestic visitors received. The first location was the capital city of Tehran, and the second was the city of Chaloos, located in the north of Iran in the Caspian Sea region. Chaloos was selected because of its location in one of the highly popular sun, sea, and sand tourism zones; the area is also popular for its green and lush environment. In addition, Chaloos plays the role of a "pressured area", a rural region within the sphere of influence of major urban areas (i.e. Tehran), where the countryside is used for day trip recreation.

Non-probability sampling was administered based on a mode of sampling known as "convenience" sampling (Bryman, 2004). The responses were analysed using SPSS, version 13.0. Of 500 self-completed questionnaires, 400 were distributed in various travel agencies in Tehran, and the remainder was filled out by respondents in Chaloos, where tourists were targeted in different hotels. A total of 483 valid questionnaires were collected and analysed; each questionnaire form contained 42 questions in three sections. The first 33 survey questions were designed to measure responses based on a five-point Likert scale (mean measured on a five-point scale ranging from 1 = strongly agree to 5 = strongly disagree). The second section was composed of four questions regarding demographic information to identify age, gender, education, and frequency of travel. The third section of the questionnaire was composed of five items to clarify the purpose of travel, seasonality, length of stay, availability of information prior to trip, and satisfaction with the facilities. Eighty-two per cent of the respondents were male. Over 30% of the respondents were under 40 years of age. The majority of the respondents were university graduates (40%), and 36.6% of the respondents travelled twice a year (frequency of travel pattern).

Table 1 contains descriptive statistics for 33 attitudinal statements about respondents' perception regarding overall problems and bottlenecks that travelling in Iran is associated with. Respondents also expressed their views regarding the benefits and impacts of domestic tourism. Respondents were asked to express their views regarding the social, environmental, and cross-cultural issues, and public sector's role in the context of domestic tourism. Table 1 shows the mean values, frequency, cumulative percentage, and standard deviation for the "strongly agree = 1" level that is extracted from the main descriptive statistics (see Table 1). As shown in Table 1, it was found that respondents agreed strongly with the attitudinal statements (grand mean value = 1.65). In the meantime, mean score and ranking of 10 attitudinal statements are also shown in Table 2, in which respondents believed that issues contained in these statements are the most pressing items relevant to domestic tourism. The ranking exercise revealed that domestic tourists were critical of sanitation and quality (mean value = 1.49), and highly critical with transportation facilities and overall means of transportation (mean value = 1.49); they were also critical of government's apathy towards improving and facilitating domestic tourism (mean value = 1.50). They have also valued the cultural and historical heritage as significant motivating factors (mean value = 1.51). They openly expressed the lack of protection of heritage sites (mean value = 1.52). They were also unhappy with the quality of food outlets and rest areas (mean value = 1.53). They believed that domestic tourism can be a catalyst to development in rural areas (mean value = 1.54). Contribution of domestic tourism to the social welfare (mean value = 1.57), and cross-cultural understanding (mean value = 1.62) were the other items that respondents emphasized. In this section of the methodology, five more questions were applied, which are considered non-Likert type; they addressed pattern of travel, satisfaction with the facilities (i.e. facilitating services), duration of the trip, seasonality (i.e. seasons in which travelling frequency is highest), and availability of information prior to the trip.

Table 1 Descriptive statistics of the Likert-scale result for responses (strongly agree scale).

Questions	Strongly agree (SA)	Frequency	Cumulative (%)	Mean	SD
1	SA	255	56.2	1.64	0.888
2	SA	268	55.7	1.51	0.639
3	SA	250	52.6	1.57	0.691
4	SA	210	43.5	1.69	0.722
5	SA	226	47.6	1.66	0.724
6	SA	226	47.0	1.62	0.673
7	SA	229	48.0	1.67	0.763
8	SA	239	50.1	1.61	0.709
9	SA	218	45.5	1.69	0.763
10	SA	187	39.0	1.80	0.797
11	SA	266	56.2	1.52	0.695
12	SA	204	43.4	1.74	0.768
13	SA	222	46.3	1.66	0.700
14	SA	261	54.7	1.50	0.600
15	SA	254	53.4	1.54	0.672
16	SA	218	45.7	1.64	0.692
17	SA	242	50.6	1.58	0.683
18	SA	253	53.4	1.54	0.663
19	SA	222	46.0	1.65	0.723
20	SA	267	55.4	1.53	0.697
21	SA	278	57.8	1.49	0.665
22	SA	283	59.2	1.49	0.687
23	SA	208	43.3	1.66	0.698
24	SA	190	39.8	1.80	0.796
25	SA	233	48.4	1.67	0.799
26	SA	184	38.3	1.73	0.686
27	SA	147	30.7	1.94	0.835
28	SA	198	41.9	1.68	0.671
29	SA	194	40.7	1.70	0.709
30	SA	186	39.1	1.74	0.711
31	SA	186	39.2	1.74	0.707
32	SA	184	38.6	1.95	0.991
33	SA	222	46.4	1.69	0.793

Table 2 Mean score and ranking of 10 attitudinal statements.

	Attitudinal statements	Mean	SD	Frequency	Cumulative (%)	Rank
1	Overall quality of sanitation can encourage domestic travelling	1.49	0.665	255	56.1	1
2	Lack of adequate and safe transportation system is an obstacle to domestic travelling	1.49	0.687	283	59.2	1
3	Public sector should actively involve in focusing on positive aspects of domestic tourism	1.50	0.600	204	43.4	2
4	The historical places of Iran are one of the main motivations for domestic travellers	1.51	0.639	268	55.7	3
5	The destruction of historical places is equal to the destruction of the national culture	1.52	0.695	266	56.2	4
6	Quality of food, food outlets, and rest-areas can encourage domestic travelling	1.53	0.697	253	53.4	5
7	Domestic tourism can help to improve the rural economy	1.54	0.663	210	43.5	6
8	Travelling (domestic tourism) leads to better social health	1.57	0.691	267	55.4	7
9	Maintaining historical places results in expansion of tourism industry	1.61	0.709	229	48	8
10	Domestic tourism contributes to appreciation of different cultures	1.62	0.673	226	47	9

CONCLUSION AND IMPLICATIONS

This study has focused on domestic tourism in the context of the Iranian political economy. And it has reflected the tenable nature of domestic tourism. In this regard, Iran's political economy manifests a lack of cohesiveness among numerous political and economic cliques (both formal and informal), who do not realize the advantages of domestic tourism as a potential sustainable mode of tourism (Ghaderi, 2011; Seckelmann, 2002). The overarching problem identified in this study is the policy-makers' suspicions of international tourism, which, to some extent, is understandable; however, this attitude need not extend to domestic tourism, which is a significant developmental tool.

Iran's case is unique and complex. One can easily witness the clashes between the ideas of modernity and traditionalism that make achieving reconciliation cumbersome (Jahanbegloo, 2004). In the meantime, the dynamism of Iranian society has enough space to pursue a rational planning process towards a formidable and sustainable domestic tourism system. This study has tried to fulfil the need to theorize the nature of power, conflict, development and political agency in the context of domestic tourism, and the lack of a sustainable planning instrument. For instance, domestic tourism and its growth in Iran have become a challenge, where citizens are demanding recreation and leisure as their human right, while officials perceive it as luxurious decadence associated with the western way of life.

This study examined the anatomy of the domestic tourism project as a wakeup call to the public sector, not only to fulfil the demands of the tourism market within Iran but also as a way out for the policy-makers who are suspicious of tourism and complacent regarding the tenacity of Iranians in travelling. Therefore, a domestic tourism project should be analysed and justified on three grounds. First, domestic tourism has been identified as an activity that can be immune to the vagaries to which global tourism is subjected in the context of the global economic structure. This perspective, under the culture-ideology of consumerism, is highly critical of mass tourism as an active dimension of the global accumulation of wealth and controlled by global tourism, which is motivated by profit through the capitalist production of mass tourism. Proponents of this theory argue that, with respect to production and consumption as the key to growing profits, the global structure of tourism contradicts the implementation of sustainability principles in relation to ecology, society, and economics (Azarya, 2004; Britton, 1991; Buckley, 2012; Higgins-Desbiolles, 2009; Sklair, 2002).

Second, Iranian domestic tourism and its system definition were examined through interviews with the officials who are supposed to agree upon and be involved in a framework to govern such a system. Without explicit recognition and a common understanding among formal sectors, it would be almost impossible to agree on the entity for which the strategy is to be developed. The result of the qualitative approach revealed that the absence of an explicit philosophy resulted in a lack of development of a coherent domestic tourism policy (Ritchie & Crouch, 2003). Notwithstanding Iran's dynamic religious tourism, the domestic tourism system should embrace and facilitate numerous compatible activities with existing religious practices without jeopardizing the sacred traditions.

Such a system must consider several fundamental issues. First, domestic tourism will save Iran's billions of dollars that leave the country through outbound tourism. This is a case where the country is experiencing a shortage of hard currency and economic hardship (Gladstone, 2012). Second, in a multi-ethnic Iran with regional disparities and a gap between rural and urban areas, domestic tourism has the potential to contribute to cross-cultural understanding and rural development (Brouder, 2012; Sheykhi, 2009) that can strengthen communal bonds and modify negative stereotypes. Third, domestic tourism, if actively supported by and involved with the public sector, can lay out the groundwork for establishing strong domestic tourism now and viable international tourism in the future (Ritchie & Crouch, 2003).

The study has drawn upon the classic concept of sustainable development, which has been accepted as a strategic framework for sustainable tourism. The policy-makers in Iran's NTO and beyond need to overcome certain challenges including accommodating citizen's recreational needs in the context of an Islamic culture. Domestic tourism, contrary to international mass tourism, is a national system that is not tied to the global pattern of capitalism. This study suggests that public policy-makers in Iran should embrace domestic tourism through an ADTM in the context of Iran's political economy. The proposed model is a framework that would help to reconcile the policy-makers' suspicious attitude towards international mass tourism, to identify the bottlenecks that hamper the tourism system in Iran, and to accommodate the population's recreational needs. The evidence presented here demonstrates the need to take into consideration Iran's ethnic variations, vagaries of global tourism structure, and Iran's political and ideological environment to define, shape, and restructure policies for sustainable domestic tourism.

NOTES

[1.] Email: hasan.kilic@emu.edu.tr

[2.] Email: nono_Z1169@yahoo.com

REFERENCES

1. Azarya, V. (2004). Globalization and international tourism in developing countries: Marginality as a commercial commodity. Current Sociology, 52, 949–967. [CrossRef]

2. Beck, U. (2002). The cosmopolitan society and its enemies. Theory, Culture and Society, 19, 17–44.

3. Bianchi, V. R. (2004). Tourism restructuring and the politics of sustainability: A critical view from the European periphery (The Canary Islands). Journal of Sustainable Tourism, 12, 495–529.

4. Britton, S. G. (1991). Tourism, capital and place: Towards a critical geography of tourism. Environment and Planning D, 9, 451–478.

5. Brouder, P. (2012). Creative outposts: Tourism's place in rural innovation. Tourism Planning and Development, 9, 383–396.

6. Bryman, A. (2004). Social research methods. New York: Oxford University Press.

7. Buckley, R. (2012). Sustainable tourism: Research and reality. Annals of Tourism Research, 39, 528–546.

8. Burns, P. (1999). Paradoxes in planning: Tourism elitism or brutalism?Annals of Tourism Research, 26, 329–348.

9. Burns, P. (2004). Tourism planning: A third way. Annals of Tourism Research, 31, 24–43.

10. Carvalho, G. O. (2001). Sustainable development: Is it achievable within the existing international political economy context?Sustainable Development, 9, 61–73.

11. Chao, C. C., Hazari, B. R., & Sgro, P. M. (2004). Tourism, globalization, social externalities, and domestic welfare. Research in International Business and Finance, 18(2), 141–149.

12. Dauvergne, P. (1997). Shadows in the forest: Japan and the politics of timber in Southeast Asia. Boston: MIT Press.

13. Demir, C. (2004). A profile of Turkish tourism. International Journal of Contemporary Tourism Management, 16, 325–328.

14. Dodds, R. (2012). ustainable tourism: A hope or a necessity? The case of Tofino, British Columbia, Canada. Journal of Sustainable Development, 5, 54–64.

15. Driml, S., & Common, M. (1996). Ecological economics criteria for sustainable tourism: Application to the great barrier reef and wet tropics world heritage areas, Australia. Journal of Sustainable Tourism, 4, 3–16.

16. Fetterman, D. M. (2010). Ethnography: Step-by-step (3rd ed.). London: Sage.

17. Fletcher, R. (2011). Sustaining tourism, sustaining capitalism? The tourism industry's role in global capitalist expansion. Tourism Geographies, 13, 443–461.

18. Ghaderi, Z. (2011). Domestic tourism in Iran. Anatolia, 22, 278–281.

19. Gladstone, R. (2012, October 16). Iran sanction may cut supply of currency. *The New York Times*. Retrieved from http://www.nytimes.com/2012/10/17/world/middleeast/irans-supply-of-currency-may-be-at-risk-in-sanctions.html?_r = 0.

20. Hall, M. C. (2008). Tourism planning: Policies, processes and relationships. London: Pearson.

21. Hall, M. C. (2011). A typology of governance and its implications for tourism policy analysis. Journal of Sustainable Tourism, 19(4–5), 437–457.

22. Hardt, M., & Negri, A. (2004). Multitude: Wars and democracy in the age of empire. New York: Penguin.

23. Hanna, K. S. (2005). Planning for sustainability. Journal of the American Planning Association, 71, 27–40.

24. Higgins-Desbiolles, F. (2008). Justice tourism and alternative globalization. Journal of Sustainable Tourism, 16, 345–364.

25. Higgins-Desbiolles, F. (2009). International solidarity movement: A case study in volunteer tourism for justice. Annals of Leisure Research, 12, 333–349.

26. Higgins-Desbiolles, F. (2010). The elusiveness of sustainability in tourism: The culture-ideology of consumerism and its implications. Tourism and Hospitality Research, 10, 116–129.

27. Higgins-Desbiolles, F. (2012). The hotel bauen's challenge to cannibalizing capitalism. Annals of Tourism Research, 39, 620–640.

28. Higgins-Desbiolles, F., & Whyte, K. P. (2013). No high hopes for hopeful tourism: A critical comment. Annals of Tourism Research, 40, 428–433.

29. Hosseini, H. (2003). Why development is more complex than growth: Clarifying some confusions. Review of Social Economy, 61, 92–110.

30. Hunter, C. (1997). Sustainable tourism as an adaptive paradigm. Annals of Tourism Research, 24, 850–867.

31. Jahanbegloo, R. (Ed.). (2004). Iran: Between tradition and modernity. Oxford: Lexington Books.

32. Larson, L. R., & Poudyal, N. C. (2012). Developing sustainable tourism through adaptive resource management: A case study of Machu Picchu, Peru. Journal of Sustainable Tourism, 20, 917–938.

33. Lash, S., & Urry, J. (1994). Economies of sign and space. London: Sage Publications.

34. Loomis, T. M. (2000). Indigenous populations and sustainable development: Building on indigenous approaches to holistic, self-determined development. World Development, 28, 893–910.

35. Martinez-Alier, J., Pascual, U., Viven, F. D., & Zaccai, E. (2010). Sustainable de-growth: Mapping the context, criticisms and future prospects of and emerging paradigm. Ecological Economics, 69, 1741–1747.

36. Mehryar, A. H., & Ahmad-Nia, S. (2004). *Age-structural transition in Iran: Short and long term consequences of drastic fertility swings during the final decades of twentieth century*. Retrieved from http://www.cicred.org/Eng/Seminars/Details/Seminars/Popwaves/PopwavesMehryarAhmad-Nia.pdf.

37. Mitchell, J. V. (2002). A new political economy of oil. The Quarterly Review of Economics and Finance, 42, 251–272.

38. Moufakkir, O. (2010). Peace through domestic tourism and tourism rights: Inclusion of Muslim ethnic minorities in the context of social tourism. The Journal of Tourism and Peace Research, 1, 42–59.

39. Mura, P., & Tavakoli, R. (2012). Tourism and social capital in Malaysia. Current Issues in Tourism, 10.1080/13683500.2012718320.

40. O'Gorman, K. D., Baum, T., & McLellan, L. R. (2007). Tourism. In Iran: Central control and indigeneity. In R.Butler & T.Hinch (Eds.), Tourism and indigenous peoples: Issues and implications (pp. 251–264). Burlington: Butterworth-Heinemann.

41. Papatheodorou, A., Rosselló, J., & Xiao, H. (2010). Global economic crisis and tourism: Consequences and perspectives. Journal of Travel Research, 49, 39–45.

42. Ritchie, J. R. B., & Crouch, G. I. (2003). The competitive destination. Wellingford: CABI

43. Roudometof, V. (2005). Transnationalism, cosmopolitanism and glocalization. Current Sociology, 53(1), 113–135.

44. Saarinen, J. (2006). Traditions of sustainability in tourism studies. Annals of Tourism Research, 33, 1121–1140.

45. Schilcher, D. (2007). Growth versus equity: The continuum of pro-poor tourism and neoliberal governance. Current Issues in Tourism, 10(2–3), 166–193.

46. Seckelmann, A. (2002). Domestic tourism—A chance for regional development in Turkey?Tourism Management, 23, 85–92.

47. Sheykhi, M. T. (2004). Globalizing influences on leisure: A perspective from Iran. World Leisure Journal, 46, 59–67.

48. Sheykhi, M. T. (2009). Domestic tourism in Iran. Tourismos: An International Multidisciplinary Journal of Tourism, 4, 109–123.

49. Sklair, L. (1999). Competing conceptions of globalization. Journal of World-Systems Research, 2, 143–163.

50. Sklair, L. (2002). Globalization: Capitalism and its alternatives (3rd ed.). Oxford: Oxford University Press.

51. Sklair, L. (2010). Iconic architecture and the culture-ideology of consumerism. Theory, Culture & Society, 27, 135–159.

52. Tribe, J., & Airey, D. (2007). A review of tourism research. In J.Tribe & D.Airey (Eds.), Developments in tourism research (pp. 3–16). Amsterdam: Elsevier.

53. Trochim, W. M. (2006). *Research method: Concise knowledge base.* Retrieved from http://www.socialresearchmethods.net/kb/qualmeth.php.

54. Wheeler, B. (2012). Sustainable mass tourism: More smudge than nudge – The canard continues. In T. V.Singh (Ed.), The critical debates in tourism (pp. 39–43). Bristol: Channel View Publications.

55. World Commission on Environment and Development. (1987). Our common future. Oxford: Oxford University Press.

56. World Tourism Organization. (1995). Collection of domestic tourism statistics. Madrid: World Tourism Organization.

57. Wu, B., Zhu, H., & Xu, X. (2000). Trends in China's domestic tourism development at the turn of the century. International Journal of Contemporary Hospitality Management, 12(5), 296–299.

Chapter 5

Measurement of Restaurant Manager Expectations toward *Halal* Certification Using Factor and Cluster Analysis

Sharifah Zannierah Syed Marzuki[a*], Colin Michael Hall[b], Paul William Ballantine[b]

[a]Faculty of Business Management, Universiti Teknologi MARA, Lintasan Semarak 26400, Bandar Tun Abdul Razak, Pahang, Malaysia

[b]Department of Management, College of Business and Economics, University of Canterbury, Christchurch, New Zealand

ABSTRACT

Halal certification is very important to the restaurant industry in Malaysia but the actual implementation of its standards is crucial for the comprehension of those who are unfamiliar with Islamic dietary rules. This paper determines whether there is a difference between Muslim and non Muslim restaurant manager towards *halal* certification. A total of 2080 questionnaires were administered nationwide using mail survey and received 643 returned and completed responses. Factor and cluster analyses were employed in this study to further examine the data. In factor analysis, seven factor groups were produced whereas the cluster analysis had revealed that the respondents were divided into three distinct groups within the sample. The first group was "Certification Advocates" that consists of Muslim respondents. The second group was known as "Certification

Ambivalents" that comprises of Chinese participants who were mostly Buddhists and the third group was labelled as "Compliance Cost Focused". Overall restaurant managers in Malaysia believed that having a *halal* certificate is an advantage as the market for *halal* food is rising and consumers are demanding for *halal* certified food products including those available at eating outlets.

INTRODUCTION

Halal or permitted food that is according to Islamic dietary laws is based on the guidelines from the Al-Quran (Muslim Holy Book) and *hadiths* (books that recorded the sayings and practices of the Prophet Muhammad , peace be upon Him). *Halal* certification of a restaurant represents the status of the food being offered at an eating premise is in accordance with the Muslims dietary rules. Furthermore, the *halal* certification means not only are the food served *halal* but the entire process would be *halal* as well. Therefore, *halal* certification is an important aspect in the Malaysian food service industry. Mohd Yusoff (2004) claimed that *halal* certification provides some benefits. One of them is confidence as people can make an informed choice of their purchase as well as it can affect the mentality of potential customers (Mohamed Nasir & Pereira, 2008). This is supported by Shafie and Othman (2006) that emphasized on the importance of having a producer's guarantee of claim. Here, they reasoned that having peace of mind initially came from trust in order to attract specifically the Muslim customers to come and dine at restaurants.

The opposite of *halal* is *haram* and it is essential for the industry players in the restaurant sector to avoid offering prohibited food and drinks like pork and alcohol to be served in their eating premise. In order to follow the demands of being *halal* the entire food supply chain must follow the *halal* standard including the slaughtering of animals, storage, display and preparation. Thus, the standard makes compulsory the requirement to physically separate *halal* from *haram* food and products.

There is a dearth of academic writing on *halal* foods in English with the possible exception of some specific works on *halal* meat and slaughtering (for example, Regenstein et al., 2003a; Riaz & Chaudry, 2004). It is therefore hoped that this paper will not only make a contribution to the specific focus on the expectations that restaurant managers have of *halal* certification for restaurants but also some of the wider issues surrounding *halal* food with respect to awareness, integrity and significance.

MALAYSIAN *HALAL* CERTIFICATION

The certification of *halal* food has been formally conducted throughout Malaysia since 1982. In addition to religious reasons much of the support for the process has arisen due to the enormous economic potential of worldwide demand for *halal* food and the government's interest in developing the country as a global *halal* hub. The National SME Development Council (2006) reported that the Codex Alimentarius Commission, which is responsible under the United Nations for international regulation of food preparation, has cited Malaysia as a good example in terms of the *halal* food certification process.

With that, the Malaysian model has therefore come to be regarded as a potential role model for the development of *halal* food quality assurance processes internationally (JAKIM, 2010). Establishments of the *halal* standard is related to *Shariah* requirements where *halal* food certification refers to the examination of food processes, from the preparation, slaughtering, ingredients used, cleaning, handling and processing, right down to transportation and distribution (Mohd Yusoff, 2004).

Malaysia has developed its industrial relations with regards to *halal* food through such agencies like JAKIM, Department of Standards Malaysia, Institute of Islamic Understanding Malaysia and Malaysian Institute of Industrial Research and Standard (SIRIM). As a result, a comprehensive *Halal* Food Standard called MS1500:2004 has been developed (Abdul Talib et al., 2008). The Malaysian Standard MS 1500 General Guidelines on the Production, Preparation, Handling and Storage of *Halal* Food prescribes the practical guidelines for the food industry on the preparation and handling of *halal* food and serve as a basic requirement for food product and food trade or business in Malaysia (Mohd Daud, 2004). It can also be used together with MS1480:1999 Food Safety According to Hazard Analysis and Critical Control Point (HACCP) and MS1514:2001 General Principles of Food Hygiene that set out the necessary hygiene conditions for producing safe food from primary production to the final consumer, which is suitable for consumption (Abdul Talib et al., 2008).

The Malaysian Standard MS 1500 General Guidelines on the Production, Preparation, Handling and Storage of *Halal* Food prescribes the practical guidelines for the food industry on *halal* conformance. In addition the Hazard Analysis Critical Control Point (HACCP) and the *halal* system will work together to ensure that the product is safe and can be consumed by anyone including non-Muslims. In Malaysia these various standards set the regulatory context for restaurants which must ensure that in order to be *halal* compliant they must be able to account for all aspects of the food process from farm to plate: the segments from sourcing the raw material, handling, processing equipment, processing aids, storing, transporting, preparing and delivering to the patrons.

Apart from having *halal* certification to suit the religious dietary restriction of the Muslims, restaurant managers felt that the status also provides commercial value (Mohamed Nasir & Pereira, 2008). A point raised by Al-Harran and Low (2008) with respect to *halal* marketing is that, it is very much like marketing for any other products and the 4Ps of the marketing mix should be applicable. However, it is also an integrating process where the Muslim and non-Muslim patrons can dine together in a multicultural country like Malaysia. In addition, *halal* restaurant operators must understand the *halal* concept clearly as Muslims in the country have become more religious or *halal* food concerns have increased in terms of their consumption habits (Al-Nahdi et al., 2009). Additionally, Islamic or *halal* tourism has become a new trend in the tourism industry and also a powerful force especially within the Middle East with outstanding prospects. In another aspect, as the demand for *halal* food is increasing, it is found that researchers are unaware of this significant market due to limited surveys and literature with respect to *halal* certification (Wan Hassan & Hall, 2003).

METHODOLOGY

This survey adopted mail survey to distribute 2080 questionnaires to *halal* certified, *halal* claimant and non *halal* restaurant managers in Malaysia. A systematic sampling approach was chosen after gaining the list of restaurants from the Companies Commission of Malaysia. The questionnaire comprised of five different sections and 75 items for the respondents to answer. As previous literature on restaurant manager and *halal* restaurants is extremely uncommon, thus the questionnaire for this study was self developed. A five-point Likert scale was employed and respondents were required to give a rating between 1 = Strongly disagree and 5 = Strongly agree. The questionnaire was written in English in order to ensure better understanding among respondents and to avoid misinterpretation of terms and meanings if translated in Bahasa Melayu (Malay Language). The response for this study was 643 and indicated a 31% response rate.

Data were then entered using Statistical Package for the Social Sciences (SPSS) version 18.0 in order to conduct factor and cluster analyses. Factor analysis is about variables (Mulaik, 2010) and it is used to gather information about the interrelationships among a set of variables (Pallant, 2007). Then, reliability testing is conducted before cluster analysis took place. The cluster analysis was then performed to identify either the respondents have similar expectations or not.

RESULTS

Demographic Analysis

A total of 643 restaurant managers have participated in the mail survey and most of them belong to the age group of 21-30 years old. The number of male respondents was higher than the female respondents with 397 to 246 respectively. As indicated in Table 1, the majority of participants were Muslim (61.59%), followed by Buddhist (24.73%), Christian (5.75%), Hindu (4.04%) and from other religions (3.89%). It showed that the number of Malay restaurant managers who responded to the mail survey has the highest percentage with 55.83% and the least from other religions with 9.02%. Most of the respondents involved in this study possessed the academic qualification level of secondary school (23.95%), Diploma (33.44%) and Bachelor's Degree (19.60%). The rest of the respondents held Masters Degree and others at 2.64% and 7.93% respectively.

Table 1. Summary of Personal Demographics

Personal Demographic Variables	*Categories*	*Frequency*	*%*
Age	Below 21	32	4.98
	21-30	339	52.72
	31-40	187	29.08
	41-50	61	9.49
	51-60	20	3.11
	61 and above	4	0.62
	Total	643	100%
Gender	Male	397	61.74
	Female	246	38.26
	Total	643	100%
Religion	Buddhist	159	24.73
	Christian	37	5.75
	Hindu	26	4.04
	Muslim	396	61.59
	Other	25	3.89
	Total	643	100%
Race	Chinese	189	29.39
	Indian	37	5.75
	Malay	359	55.83
	Other	58	9.02
	Total	643	100%
Educational Background	Doctor of Philosophy	0	0
	Masters Degree	17	2.64
	Bachelor Degree	126	19.60
	Diploma	215	33.44
	Certificate	80	12.44
	Secondary School	154	23.95
	Other	51	7.93
	Total	643	100%

Factor Analysis

Hutcheson and Sofroniou (1999) highlighted that the identification of factors is important. They reasoned that, it could provide useful theoretical insights into underlying relationships and patterns in the data. Thus, it took a large set of variables and looked for a way the data may be reduced or summarized using a smaller set of factors or components (Pallant, 2007). In another aspect, the primary purpose of factor analysis is to define the underlying structure in a data matrix (Hair et al., 2006). According to Pallant (2007) as well as Grimm and Yarnold (2005), a larger sample size is better and more reliable. As this study has a sample size of 643, the use of this technique is appropriate.

This section explains the result of factor analysis in understanding the expectations of restaurant managers toward *halal* certification in Malaysia. This study is very significant as there are no other studies that concentrated particularly on restaurant managers in relation to *halal* certification. The Kaiser- Meyer-Olkin (KMO) measure of sampling adequacy is 0.92, exceeding the recommended value of 0.50 by Kaiser (1974) where this explains that the result falls under great value. In addition, the Bartlett's Test of Sphericity reached statistical significance of 0.00 which indicate the χ^2 = 14610.29. Measure of sampling adequacy for individual variables from the Anti-Image Correlation Matrix ranges from 0.63 to 0.95. The responses to the survey form were subject to a Principal Components Analysis with a Varimax rotation and the results of the analysis appear in Appendix A.

Using Kaiser's criterion of retaining factors with eigenvalues greater than one, seven interpretable factors were obtained from the analysis. This was then compared to the scree plot, a graph of each eigenvalue (Y-axis) against the factor related with (X-axis) as proposed by Catell (1966). By looking at the point of inflection of the scree plot, it can be gathered that the data has seven underlying factors. Thus, the analysis produced seven factors where the total variance explained was 77% with all eigenvalues being over one. The title given to each factor is named and described as in Appendix A. With cut off value of 0.65 for inclusion of a variable in interpretation of a factor, 14 of 37 variables did not load on any factor.

These 14 items are identified to have values below 0.3. Low communality values of less than 0.3 indicate that the variables do not fit well with the variables in its component and should be removed from the scale. As a rule of thumb, only variables with loadings of 0.32 and above are interpreted (Tabachnick & Fidell, 2007). Therefore, the greater the loading, the more the variable is a pure measure of the factor. This criterion for the significance of factor loadings is agreed by Hair et al., (1995) that considered the loadings of 0.30 as significant only for sample sizes of 350 or greater. Considering both suggestions, factor loadings of 0.65 is selected as this study has a large sample size of 643. Furthermore, the choice of cut off value for size of loading to be interpreted is upon the researcher to decide (Tabachnick & Fidell, 2007).

By using factor analysis, it is believed that the integration of the responses allows the author to answer questions of substantial importance. Furthermore, due to the complexity of areas that covers data from a large number of perspectives (expectations of your own restaurant, expectations toward *halal* certification in general and expectations toward *halal* certification in Malaysia), factor analysis is seen to be a tool for cross validation and acts a link between further discussion on cross tabulation in order to gain a more complete understanding.

- Factor 1 consisted of "Market Signals" that indicates *halal* certification attributes such as healthy food, taste genuity, food quality, authenticity, safety and hygiene. Since the highest variance explained is 33.39%, it proved that respondents expected that *halal* certification represents important factors on market signals.
- Factor 2 showed the "Islamic Attributes". This dimension had four components that integrate the Muslims with *halal* food such as *halal* certification increases confidence among Muslim customers, signifies conformance to *halal* standards, increases demand from Muslim customers and signifies all food sources are *halal* certified. The variance explained is 13.40% and thus represents essential factors expected by the respondents toward *halal* certification at restaurants relating to Islamic attributes.
- Factor 3 highlighted on "Marketing Benefits" and in this grouping, *halal* certification acts as a marketing strategy tool, attracts the tourism market, a trademark establishment and gives a competitive advantage over non-certified restaurants. These items in this factor explained 9.31% of the variance.
- Factor 4 is "Cost of Certification" that is related to tangible and intangible costs such as applicants will have to pay a great amount of money, the involvement of lengthy time and existence of bureaucracy in getting the approval. As indicated in Appendix A this dimension explained 6.58% of the variance.

- Factor 5 is on "Cost of Compliance" where two components related to issues on *halal* certification such as the *halal* standard is difficult to follow as well as to maintain it. The variance explained was 5.35%.
- Factor 6 indicated "Non-Islamic Attributes" that highlighted on *halal* certification increases demand and confidence from non-Muslim customers that gives a variance explained of 4.84%.
- Factor 7 is related to "Local Awareness" where the dimension consisted of familiarity of the local people towards *halal* certification and gaining acceptance from them. It shows a variance explained of 4.39%.

4.3. Reliability Analysis

As mentioned earlier, an eigenvalue of more than 1.0 was used as a determinant criterion for each factor in the factor analysis. In addition, internal consistency analysis (Cronbach α) was applied to confirm the reliability of each research factor where validity and reliability tests are essential for testing the goodness of measure. According to Sekaran (2000), the closer the Cronbach's alpha value is to one, the higher the internal consistent reliability. Thus, the results in Appendix A showed that most of the factor loading exceeded 0.6 (Nunnally, 1978) as the significant level of convergent validity. The Cronbach's α values for all dimensions range from 0.84 to 0.92 which exceeds the minimum α of 0.6 (Hair et al., 2006), thus the constructs measure are deemed reliable. In another context, the mean scores in Table 2 explained the importance level of each factor. Again, market signals showed the highest mean score of 4.57 and the lowest score was cost of compliance with 2.71.

4.4 Cluster Analysis

As this study is essentially exploratory, cluster analysis was further employed to determine if respondents hold common attitudes and expectations about halal certification. Thus, this cluster analysis was used to uncover any similarities that might be used to help identify distinct groups within the sample using the seven factors identified in the previous section. Cluster analysis is a technique that almost constantly produces groupings that are significantly different from each other (Ryan and Huyton, 2000). Seven dimensions from factor analysis have been used to perform the clustering technique using SPSS version 18.0 to provide better definition of the clusters. Previously, in order to see each dimension's internal consistency, the coefficient alpha of reliability was computed. The data from expectations toward halal certification in general and expectations toward halal certification in Malaysia were used to segment the respondents for a better understanding of their expectations.

Table 2. Means, Standard Deviations and Reliabilities of Measures Used.

Factor	*Mean*	*SD*	*Cronbach's α*
Market Signals	4.57	.51	.92
Islamic Attributes	4.38	.67	.87
Marketing Benefits	3.89	.84	.84
Cost of Certification	3.31	.90	.86
Cost of Compliance	2.71	.94	.92
Non-Islamic Attributes	3.58	.96	.86
Local Awareness	4.07	.78	.84

The seven dimensions identified by the factor analysis provided the data for cluster analysis instead of the original ratings on variables. According to Mueller and Kaufmann (2001), cluster analysis is commonly used in for the classification of objects. Furthermore, this analysis covers a great variety of methods and algorithms for identifying groups of much related objects. On the basis of similarity, the cluster should be as internally and externally homogeneous in order to distinguish from other clusters. Restaurant managers who are the respondents in this survey were segmented in order to group homogeneous expectations toward *halal* certification. Given the sample size of 643, a non hierarchical clustering approach (K- means) in which the respondents are moved between clusters in order to minimize the variability within clusters and maximize the variability between clusters was adopted. Identification of clusters in this study is based on the guidelines by Aldenderfer and Blashfield (1985) for further investigation.

A visual inspection was performed of the horizontal icicle dendogram and the sudden jumps in the algorithm schedule. The results suggested that three solutions might be suitable for further analysis because it provided the greatest difference between clusters and it yielded the most interpretable results. It was anticipated that significant differences would occur in each segment. Table 3 presents the three different groups and indicates the first group "Certification Advocates" had the highest market signal, Islamic attribute, marketing benefits, non-Islamic attribute and local awareness ratings. The number of respondents in this group was 190 (29.55% of the sample) who perceived that *halal* certification contained a number of benefits.

In contrast, the second group "Certification Ambivalents" had the lowest market signal, Islamic attribute, marketing benefit, non-Islamic attribute and local awareness ratings. This group represented 251 respondents (39.03% of the sample). Next, the third group "Compliance Cost Focussed" perceived that *halal* certification as being meaningful to the restaurants but also perceived the certification process as being costly in terms of financials and time. There were 202 respondents in this group (31.42% of the sample). This study also found that there were no differences in terms of respondent age, gender or educational background. It clearly stated that in terms of religion, Muslim respondents were more likely to be "Certification Advocates" while Buddhist respondents were more likely to be "Certification Ambivalents" (χ^2 = 24.922, p = .002). Looking at race, Malay respondents were more likely to be "Certification Advocates" while Chinese respondents were more likely to be "Certification Ambivalents" (χ^2 = 19.272, p = .004).

Further comparisons were made between the groups in terms of the restaurant each respondent was referring to, when answering the questionnaire. There were no differences found in terms of either seating capacity, total number of employees, the number of years they had been working in their current position at the restaurant or the number of years they had been working in the food service industry. An interesting finding showed that "Certification Ambivalents" were more likely to work in non-*halal* restaurants while "Certification Advocates" were more likely to be working in *halal* certified restaurants (χ^2 = 29.445, p =.000).

DISCUSSION

The findings showed that respondents had high expectations toward *halal* certification where seven important factors emerged from factor analysis such as market signals, Islamic attributes, marketing benefits, cost of certification, cost of compliance, non Islamic attributes and local awareness. The highest variance explained was market signals that highlighted *halal* certification connotations. For example, it signifies healthy food, taste, food quality, authenticity, safety and hygiene. By looking at the wholesomeness concept of *halal* food, it entails every aspect according to the *Shariah* principle. As such, maintaining the standards is required to obtain good food in terms of hygiene, sanitation and safety.

Islamic attributes that appeared as the second factor is regarded as increasing the confidence and demand of Muslim customers, conforming to *halal* standards, and signifying all food sources are *halal* certified. The satisfaction based on trust increases the confidence level among the Muslim customers when deciding to visit an eating premise. It is gathered from the findings that this element of trust covers all aspects in the food supply chain as well as the dining experience. The importance of trust in this study of restaurant managers is similar to other studies on *halal* food, products and restaurants (Al-Nahdi et al., 2009; Dahalan, 2008; Wan Omar et al., 2008).

It can be suggested from this study that *halal* certification has a quality label or trademark establishment. Most of the restaurant managers agreed that with the trademark, *halal* certification acts as mode of communication specifically to the Muslim customers that the food complies with their religious dietary rules. In essence, this gives a competitive advantage over non certified *halal* restaurants. This is one of the marketing benefits that a restaurant manager can seek from *halal* certification as the results in Appendix A indicate that the majority of the respondents agreed that *halal* certification can be adopted as a marketing strategy tool. Previous researches proved that the acceptance towards *halal* foods is increasing not only among the Muslims but also among the non Muslims (Abdul Latif, 2006; Muhammad, 2007; Nik Muhammad et al., 2009). Looking at this phenomenon, the potential in this market is remarkable.

Factors on cost of certification and cost of compliance are more related to financial and non financial costs such as the fees, duration taken for Department of Islamic Development Malaysia/ Jabatan Kemajuan Islam Malaysia (JAKIM) to approve and award *halal* certification, following the *halal* standard and maintaining to the standard. The next factor on non Islamic attributes revealed that *halal* certification in a sense can increase the demand and confidence level of the non Muslim customers with a variance of .86 and .82 respectively. Muhammad (2007) and Abdul Latif (2006) reiterated that *halal* food is for all races and religions. The last factor on local awareness highlighted that *halal* certification is familiar and gaining acceptance from the local people. It is gathered that Muslim customers were seeking for *halal* food as proven from the study conducted by Al-Nahdi et al. (2009) as it would increase their intention to patronage *halal* restaurants.

The findings from the cluster analysis highlighted the different expectations of Muslim and non- Muslim restaurant managers. In essence, there was a clear difference between the Muslim and non- Muslim respondents towards *halal* certification in the survey. The result of the cluster analysis proved that the respondents were divided into three distinct groups. Cluster 1 is named "Certification Advocates" who were the Malays and Muslim restaurant managers. The members of cluster 1 had the highest market signal, Islamic attributes, marketing benefits, non-Islamic attributes and local awareness ratings. The members felt that *halal* certification had a lot of benefits where it is essential to follow the *halal* standard procedure and were confined with their daily routines in delivering *halal* values to the customers. This group appeared to be more likely working in *halal* certified restaurants. They expected that *halal* certification in restaurants was more towards gaining positive impact on customer satisfaction. They believed that maintaining the *halal* standards was the most important factor in running their business.

In comparison, cluster 2 is known as "Certification Ambivalents" comprising of Chinese respondents who were Buddhists had the lowest market signal, Islamic attributes, marketing benefits, non-Islamic attributes and local awareness ratings. In fact, they were the opposite in perceiving *halal* certification benefits compared to "Certification Advocates". These groups of respondents were more likely to be working in non-*halal* restaurants. However, findings from the interviews indicated that non-Muslim participants suggested that Islamic norms and values must be practised while operating *halal* certified restaurants. This is to preserve the Muslim customers' dietary restriction in particular and gain mutual understanding from the society in general. The managers had viewed *halal* certification in a broader yet practical perspective and with the combination of the religious aspects they noted that *halal* status encompassed several important factors for the benefits of their restaurants.

The non-Muslim respondents of *halal* certified and *halal* claimant restaurants also respected the dietary restrictions of the Muslims and worked closely to fulfill and maintain the *halal* standards. This is supported by the findings from the interviews where the participants emphasized the importance of knowledge relating to Muslim dietary prohibition and religious practices. An interesting aspect is that there are some non-Muslim restaurant managers who have much interest and are well versed in Islamic dietary laws. Their sensitivity towards *halal* food and eating environment may be one of the results of living with multiple races for centuries. The last group in the cluster was "Compliance Cost Focussed", who perceived that *halal* certification brings benefits but at the same time felt that the certification process demands financial and non-financial costs.

The application process and conformance to the *halal* standard was also a significant issue for restaurant managers. Both the survey and the interviews raised similar results on the strict processes prior to be certified and *halal* compliant after certification. Again, it can be argued that the procedures are rigid due to the fact that it is a law from the *Quran*. It is clearly stated in the Holy Book that the Islamic dietary laws permits and prohibits certain foods for Muslim consumption. Thus, there is no compromise on that matter. In addition, the issues on non-government recognized *halal* certifiers have become a serious problem and affect the integrity of related enforcement. In order to reduce and eliminate the problem, and help satisfy the concerns of some *halal* certified restaurants as to the value of the official certification process, aggressive enforcement needs to be executed. The regulatory bodies must consistently monitor these eating establishments to conform to *halal* standards at all times. Furthermore, the *Halal* Act that will commence in 2011 (Idris, 2010) is seen as a positive move in facing the challenges towards the *halal* food service industry.

Table 3. Final Cluster Centres (Mean Factor Scores for Each Cluster)

Factor	*Certification Advocates*	*Certification Ambivalents*	*Compliance Cost Focussed*
Market Signals	4.80	4.34	4.63
Islamic Attributes	4.76	3.97	4.52
Marketing Benefits	4.46	3.22	4.20
Cost of Certification	3.00	3.09	3.88
Cost of Compliance	1.73	2.65	3.72
Non-Islamic Attributes	4.06	2.86	4.04
Local Awareness	4.54	3.52	4.31
Total Number	190	251	202
Percentage	29.55%	39.03%	31.42%

CONCLUSION

Halal certification does have an impact in Malaysia, as it is one of the pioneers in introducing *halal* standards and recognition as well as becoming a global *halal* hub. In order to achieve this, the country must first have a dedicated blueprint to offer clear objective and guideline in the industry (Nik Muhammad et al., 2009). The result from the factor analysis indicated that *halal* certification comprised of several factors such as market signals, Islamic attributes, marketing benefits, cost of certification, cost of compliance, non Islamic attributes and local awareness. All of the factors have moderate to high variance.

Within this context, the cluster analysis has distinguished the respondents into three different groups. The most significant groups are the Certification Advocates that comprise of the Malays or Muslim respondents and the Certification Ambivalents that include the majority of the Buddhist and other religions, thus denotes the non Muslim respondents. Currently, Malaysia's strength is upon its certification and this study proved that the Muslim and non Muslim restaurant managers had some differences in relation to their expectations toward *halal* certification.

The result of this study has proven that *halal* certification has significant factors that relate to commercial drivers for the hospitality industry. In another perspective, respondents also have spiritual aspects like Islamic attributes that emphasize on increase demand and confidence level of the Muslim customers. As mentioned in the literature, trust does play an important role in patronizing a restaurant and for return visit (Mohamed Nasir & Pereira, 2008; Al-Nahdi et al., 2009). In realizing the government's aspiration to become a *halal* hub, restaurant managers who operate certified *halal* and *halal* claimant restaurants must be fully aware that the whole process in the premise must follow the judgments of *halal* and *haram* apart from the marketing and commercial perspectives.

REFERENCES

1. Abdul Latif, M. (2006). *Training on halal procedures and standards.* Afmaasia, http://www.afmaasia.org/malaysia_2006/Annex-4.pdf (accessed March 10, 2008).

2. Aldenderfer, M. S., & Blashfield, R. K. (1985). *Cluster Analysis.* California: Sage Publications, Inc.Al-Harran, S., & Low, P. (2008). Marketing of *halal* products: The way forward. *The Halal Journal, January/February,* 42-44.

3. Al-Nahdi, M. T. S. M., Ismail, I., Haron, H., & Islam, M. A. (2009). *Intention to patronage halal restaurants among Malaysian Muslims - An issue of halal perception.* Paper presented at the Global Business Summit Conference.

4. Cattell, R. B. (1966). The scree test for the number of factor. *Multivariate Behavioral Research*, 1, 245- 276.

5. Dahalan, Z. (2008). Kecenderungan umat Islam terhadap konsep pemakanan halal: Kajian di UiTM Pulau Pinang [The tendency of Muslims toward halal food concept: A study in UiTM Pulau Pinang]. Paper presented at the ECER Regional Conference, Kelantan, Malaysia.

6. Grimm, L. G., & Yarnold, P. R. (2005). *Reading and Understanding Multivariate Statistics*. Washington: American Psychological Association.

7. Hair, J. F., Anderson, R. E., Tatham, R. L., & Black, W. C. (1995). *Multivariate Data Analysis* (4th ed.). New Jersey: Prentice Hall.

8. Hair, J. F., Black, W. C., Babin, B. J., Anderson, R. E., & Tatham, R. L. (2006). *Multivariate Data Analysis* (6th ed.). New Jersey: Prentice Hall.

9. Hutcheson, G. D., & Sofroniou, N. (1999). *The Multivariate Social Scientist*. London: SAGE Publications Ltd.

10. Idris, A. A. (2010). *Raid shows up fake halal stickers at mall.* New Straits Times, http://www.nst.com.my/articles/Raidshowsupfake*halal*stickersatmall/Article/art_print (accessed February 1, 2011).

11. JAKIM (2010). *Halal Malaysia*. JAKIM, http://www.*halal*.gov.my/v2/ (accessed October 5, 2010). Kaiser, H. F. (1974). An index of factorial simplicity. *Psychometrika*, 39, 31-36.

12. Mohamed Nasir, K., & Pereira, A. A. (2008). Defensive dining: Notes on the public dining experiences in Singapore. *Contemporary Islam, 2*, 61-73.

13. Mohd Yusoff, H. (2004). *Halal* certification scheme. *Standard & Quality News, 11,* 4-5.

14. Mueller, H., & Kaufmann, E.L. (2001). Wellness tourism: Market analysis of a special health tourism segment and implications for the hotel industry. *Journal of Vacation Marketing*, 7(1), 5-17.

15. Muhammad, R. (2007). Re-branding *halal. The Halal Journal, May/June,* 32&34. Mulaik, S. A. (2010). *Foundations of Factor Analysis* (2nd ed.). Boca Raton: CRC Press.

16. National Small and Medium Enterprise Development Council/SME (2006). *Moving Forward. Chapter 8: potential growth areas for SMEs.* Bank Negara Malaysia,

http://www.bnm.gov.my/files/publication/sme/en/2006/chap_8.pdf (accessed July 6, 2008).

17. Nik Muhammad, N. M., Md Isa, F., & Kifli, B. C. (2009). Positioning Malaysia as *halal*-hub: Integration role of supply chain strategy and *halal* assurance system. *Asian Social Science*, 5(7), 44-52.

18. Nunnally, J. C. (1978). *Psychometric theory* (2nd ed.). New York: McGraw-Hill. Pallant, J. (2007). *SPSS Survival Manual*. Sydney: McGraw-Hill.

19. Ryan, C., & Huyton, J. (2000). Who is interested in aboriginal tourism in the northern territory, Australia? A cluster analysis. *Journal of Sustainable Tourism*, 8(1), 53-88.

20. Sekaran, U. (2000). *Research methods for business: A skill-building approach* (3rd ed). New York: John Wiley & Sons Inc.

21. Shafie, S., & Othman, M. N. (2006). *Halal certification: An international marketing issues and challenges*. CTW Congress, http://www.ctwcongress.de/ifsam/download/track13/pap00226.pdf (accessed October 8, 2007).

22. Syed Marzuki, S. Z. (2012). *Understanding the expectations of restaurant managers toward halal certification in Malaysia*. Unpublished doctoral dissertation, University of Canterbury, Christchurch, New Zealand.

23. Tabachnick, B. G., & Fidell, L. S. (2007). *Using multivariate statistics* (5th ed.). Boston: Pearson Education, Inc.

24. Wan Hassan, W. M., & Hall, C. M. (2003). The demand for *halal* food among Muslim travellers in New Zealand. In C. M. Hall, L. Sharples, R. Mitchell, N. Macionis & B. Cambourne (Eds.), *Food Tourism Around the World: Development, management and markets* (pp. 81-101). Oxford: Butterworth- Heinemann.

25. Wan Omar, W. M., Muhammad, M. Z., & Che Omar, A. (2008). *An analysis of the Muslim consumers' attitudes towards halal food products in Kelantan.* Paper presented at the ECER Regional Conference 2008, Kelantan, Malaysia.

Chapter 6

Weather and Climate Information for Tourism

D. Scott[1] *, C. Lemieux

University of Waterloo, Canada

ABSTRACT

The tourism sector is one of the largest and fastest growing global industries and is a significant contributor to national and local economies around the world. The interface between climate and tourism is multifaceted and complex, as climate represents both a vital resource to be exploited and an important limiting factor that poses risks to be managed by the tourism industry and tourists alike. All tourism destinations and operators are climate-sensitive to a degree and climate is a key influence on travel planning and the travel experience. This chapter provides a synopsis of the capacities and needs for climate services in the tourism sector, including current and emerging applications of climate services by diverse tourism end-users, and a discussion of key knowledge gaps, research and capacity-building needs and partnerships that are required to accelerate the application of climate information to manage risks to climate variability and facilitate successful adaptation to climate change.

Keywords: Tourism; outdoor recreation; sustainability; adaptation; climate services; climate sensitivity

1. INTRODUCTION

The tourism sector is one of the largest and fastest growing global industries and is a significant contributor to national and local economies around the world. Tourism represents far more than just travel for leisure and holidays. Tourism encompasses travel for education, health, religion, conventions and conferences, general business travel and visiting friends and relatives. The United Nations World Tourism Organization [1] defines tourism as including: "... the activities of persons travelling to and staying in places outside their usual environment for not more than one consecutive year for leisure, business and other purposes not related to the exercise of an activity remunerated from within the place visited." According to the UNWTO [2], international tourist arrivals have grown from just over 200 million in 1980 to 922 million in 2008. International travel is also forecast to almost double to 1.6 billion arrivals by 2020 [3]. The economic importance of the sector worldwide is demonstrated by World Travel and Tourism Council [4] estimates that in 2008 the global travel and tourism industry, contributed 9.6 per cent of global Gross Domestic Product (GDP) and

7.9 per cent of world-wide employment.

While the majority of international tourism currently occurs in developed countries, the sector is a vital contributor to the economy of many developing countries. Between 1995 and 2007, it is estimated that international tourism in emerging and developing markets grew at twice the rate of industrialized countries – by 11 per cent for Least Developed Countries and 9 per cent for other low and lower-middle income economies [5]. Visitor spending represented more than 10 per cent of national GDP in 36 developing countries in 2006 [6]. The UNWTO [7] also estimates that tourism is a primary source of foreign exchange earnings in 46 out of 50 of the world's LDCs. With the growth of tourism in developing countries, international tourism is increasingly promoted by development organizations and many governments as having an important role in contributing to the United Nations Millennium

Development Goals, particularly the alleviation of poverty in LDCs, gender equality and environmental sustainability.

While comparable global statistical data on domestic tourism are not available, its volume was estimated at around 8 billion trips worldwide in 2005, of which 4 billion were estimated to be from same-day visitors and 4 billion from overnight tourists. The shorter-term nature of much domestic tourism, particularly day trips or weekend holidays, increases the importance and relevance of nowcast and short-term forecast information for decision-making by tourists. Consequently, domestic tourism is an important consideration of this review.

The tourism sector is characterized by considerable diversity and a fragmented structure. While varied conceptualizations of the subsectors that comprise the tourism sector are used in academia and by international organizations, major components include: transportation (air lines, cruise ships, rail lines, ground coaches and taxies, for example), accommodation (hotels, apartments, youth hostels, for example), food and hospitality services (restaurants, bars and pubs, for example), travel agents and tour service operators, visitor attractions (cultural or sporting events, casinos, parks, museums, for example) and tourist focused retail or service providers (insurance, conventions, tourist equipment rentals, for example). Tourism operators differ in terms of ownership (government, non- government organizations, private businesses), size (there is a predominance of small and medium-sized enterprises in the sector, but also many international conglomerates) and purpose (for-profit or non-profit, as well as conservation, education, community development mandates). Tourism operators have also adapted to provide tourism services in every climatic zone on the planet from deserts and high mountains to the tropics and polar regions. As a consequence of this heterogeneity, there are extensive differences in the nature of climate sensitivities and abilities of tourism operators worldwide to incorporate climate services into decision-making.

Equally as diverse are the motivations and characteristics of domestic and international travellers. Major global tourism market segments include: sun and beach tourism, sports tourism, adventure tourism, nature-based tourism, cultural tourism, urban tourism, health and wellness tourism, cruises, theme parks, visiting friends and relatives and meetings and conferences. The disparate climate requirements and preferences of tourists within these major market segments, as well as between groups within each major market segment (golf, ski, and windsurfing segments of sports tourism, for example), create very different demand-side climate sensitivities within the tourism sector.

The interface between climate and tourism is multifaceted and highly complex. Figure 1 outlines the temporal scales (extreme events, seasonality, interannual variability, climate change) at which climate influences different subsectors of tourism, either directly (blue lines) or indirectly (black and red lines). Importantly, climate is but one macroscale influencing factor on the tourism system and interacts with other macroscale factors as well. The selected media headlines in Table 1 provide illustrative examples of the varied climate sensitivities in tourism supply (tourism destinations and tourism operators) and demand (tourism arrivals and travel patterns) around the world.

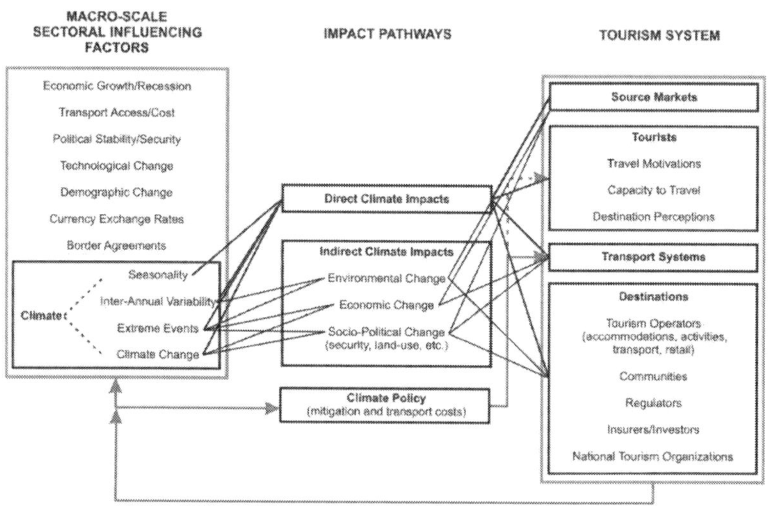

Figure 1. Climate influences on the tourism sector

Tourism destinations and tourism operators are affected by climate variability and change in a number of ways. All tourism destinations are climate-sensitive to a degree, in that they are influenced by natural seasonality in demand, are affected positively or negatively by interannual climate variability that brings heatwaves, unseasonable cold, drought or storms and heavy rain, which can affect not only tourist comfort and safety (and thereby satisfaction), but also the products that attract tourists (snow cover, coral reefs, wildlife, for example). Climate variability also influences various facets of tourism operations (water supply and quality, heating– cooling costs, snowmaking requirements, irrigation needs, pest management, and evacuations and temporary closures, for example). An international survey of 66 national tourism and meteorological organizations found that a large majority (81 per cent) felt weather and climate were major determinants of tourism in their nation [8]. Indeed some argue that climate is among the most dominant factors affecting global tourist flows [9][10].

There is a general consensus that destination image is a key determinant in destination choice (among other macroscale influencing factors, such as travel distance, time, holiday cost, and travel motivation and those set out in Figure 1) [11] and that climate is dominant attribute of destination image along with scenery and cost [12]. A review of destination image studies found that natural beauty and climate were of universal importance in defining destination attractiveness [13]. Some tourism destinations can be considered climate-dependent, in that climate is the principal resource on which tourism to the destination is predicated (for example, many Small Island Developing States).

In the same way that climate affects the destination choice of travellers it highly influences the timing of travel. Seasonal demand is one of the main defining characteristics of global tourism, and is comprised of two elements – natural and institutional seasonality [14]. Seasonal climate fluctuations at tourism destinations and at major outbound markets, particularly at high latitudes, are a key driver of tourism demand at global (Figure 2) and regional scales. When climatic resources are no longer suitable for certain tourism markets, such as ski or beach holidays, tourism operators can be forced to close seasonally (Figure 3).

Table 1. Recent media headlines of weather/climate impacts on the tourism sector

Africa	"Cape Town heat wave fans fires ahead of tourism season" (*Agence France Presse*, 9 December 2008) "Soggy weather proves bad month for tourists" (*Africa News*, October 2008) "Mozambique tourist resorts struggle to recover from cyclone" (Agence France Presse, 18 March 2008)
Americas	"Snow joke -- Ski hills are sweating, but the golf courses are rolling in green with this wacky weather" (*The Toronto Sun*, 30 November 2009) "A race to get snow in time for Olympic Games; Backup plan activated after warm temperatures plague Cypress Mountain" (*The International Herald Tribune*, 23 January 2010) "California tourism fighting fires, too" (*USA Today*, 10 July 2008) "Hurricane Ike tosses Caribbean tourism for a loop" (*Miami Herald*, 10 September 2008) "Officials try to lure tourists back to the storm-raked Caribbean" (*USA Today*, 16 September 2008) "Tourism's winter of discontent: warm weather has seasonal industries feeling the heat" (*Globe and Mail*, 4 June 2005) "Wilma deals $800 million blow to Mexican travel industry" (CNN, 27 October 2005)
Europe	"July rainfall takes toll on tourism industry" (*Irish Times*, 2 August 2008) "Weather puts a damper on tourism" (*Daily Mail*, 15 September 2007) "Devon gloomy weather forecasts keep away visitors" (*Telegraph*, 31 July 2008) "Ski resorts sweat in warm weather" (BBC News, 4 December 2006) "Tourism fears as heatwave sparks fires across Greece" (*The Guardian*, 26 July 2007) "Forget Summer: washout will wreck Brits' fun" (Daily Star, 10 May 2010)
Asia and South/East Asia	"A snow-less Srinagar disappoints tourists on New Year" (Asian News Int'l, 31 December 2007) "Snow storms costs China's tourism sector nearly one billion U.S. dollars" (Xinhua General News Service, 21 February 2008) "Freezing weather dampens during Spring Festival holiday" (BBC, 9 February 2008) "Andaman tourism dampened by tropical storm" (*The Nation*, 4 May 2007)
Oceania	"Queensland tourism hit by wild weather" (ABC Premium News, 12 February 2008) "Ongoing rain keeps Sun Coast tourists away" (ABC Regional News, 27 February 2008) "Storms battle coastal tourism" (*The Courier Mail*, 3 January 2008) "Cooling weather brings visitor lull" (*The Press*, 12 April 2008) "Weather thwarts tourism operators" (*The Press*, New Zealand, 25 January 2010) "Cyclone slams Queensland tourism operators" (*Australian Financial Review*, 24 March 2010)

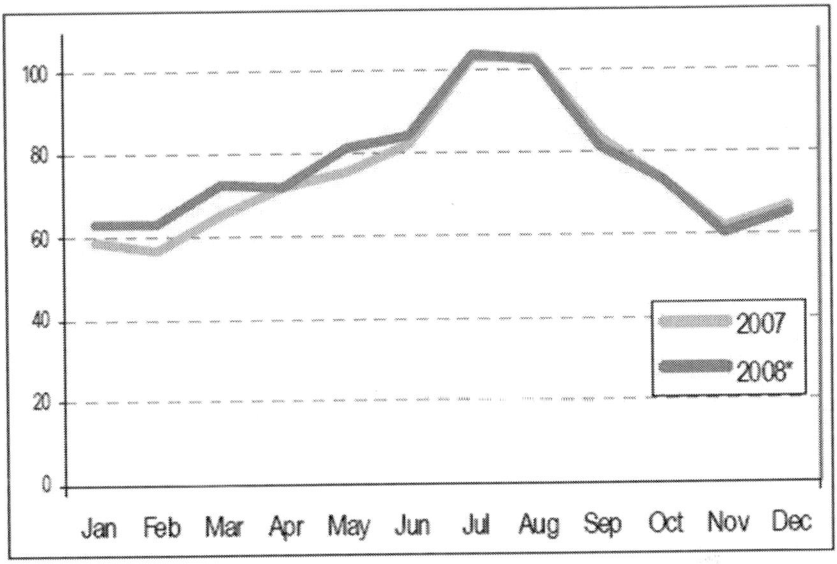

Figure 2. Monthly international tourist arrivals (millions) (Source: UNWTO [2])

Figure 3. Seasonal closure of major resorts in Greece (Photo: Daniel Scott)

Climate also has an important influence on environmental conditions that can deter tourists, including infectious disease, wildfires, insect or water-borne pests (for example, jellyfish and algae blooms) and extreme events such as tropical cyclones. In Greece, after the devastating fires of summer 2000, more than 50 per cent of all bookings from tourists for 2001 were cancelled [15]. Drought in the state of Colorado (United States) during the spring and summer of 2002 created dangerous wildfire conditions and the media coverage of major fires in some parts of the state had a significant impact on summer tourism. Visitor numbers declined by 40 per cent in some areas of Colorado and reservations in state park campgrounds dropped 30 per cent [16]. The drought also affected fishing and river-rafting tourism in the state. Anglers were restricted from fishing in many rivers because fish populations were highly stressed by low water levels and high water temperatures. Low water levels also shortened the river-rafting season substantially. Some river-rafting outfitter companies lost 40 per cent of their normal business and economic losses in the state's river rafting industry alone exceeded US$ 50 million [17][18].

Extreme weather events routinely influence tourism operators and travel decisions in regions such as the Gulf of Mexico. The economic impact of hurricanes on tourism in this region is often substantial. In the Florida Keys, the ten-day closure and clean-up following Hurricane Georges in 1998 resulted in tourism revenue losses of approximately US$ 32 million [19]. The economic impact of the four hurricanes that struck the state of Florida in 2004 was estimated to be several times larger, as the storms caused thousands of cancellations as travellers went elsewhere, and a marketing survey found that 25 per cent were also less likely to visit Florida during hurricane season in the future [20][21]. Importantly, these same extreme events in one destination had a positive impact on other parts of the tourism system, as destinations such as Arizona and California benefited from the transfer of large numbers of visitors and convention business [22]. That same year, the government of Mexico estimated that as a result of the late season Hurricane Wilma and media coverage of damage and stranded tourists, it would lose US$ 800 million in tourism revenue between October and December [23]. With 26 tropical storms and 14 hurricanes, the 2005 hurricane season was one of the most active and destructive in history, spawning three of the most intense North Atlantic storms on record, including Hurricane Katrina which caused extensive damage to the tourism infrastructure in New Orleans and coastal Mississippi, where impacts on convention business and gambling are expected for years, perhaps decades, to come [24].

The effects of the 2003 summer heatwave on tourism are also illustrative of the complexities of the climate and tourism interface. The 2003 heatwave in western Europe was exceptional by its temperatures and also its length. Tourism establishments in the Spanish beach destination of Costa Brava reportedly lost an estimated 10 per cent of guest nights during the summer season, with decreased stays at campsites most pronounced, while visitation to inland mountain destinations increased as travellers sought comfortable climatic conditions [25]. Similar shifts in tourist patterns were documented in France, with increased occupancies in destinations in northern and north-western shores and central mountains, and decreases in urban centres and southern regions. Activities, accommodations and consumption patterns were also modified, as access to some forests with high fire risk was blocked, fishing restricted, and camping and accommodations without adequate space-cooling systems became uncomfortable, while demand for accommodations with pools increased and sales of beverages and ice cream increased substantially [26].

Tourism operators in certain market segments are also profoundly affected by interannual climate variability and extremes. Climate defines the length and quality of multibillion dollar tourism seasons in different regions. Winter sports tourism, for example, is highly climate-sensitive. Figure 4 demonstrates the impact of interannual climate variability on the length of ski seasons in the ski regions of the United States. The marginal snow conditions during the record warm winter of 2001/02 (Figure 5) in North-east, South-east and Midwest regions had important impacts on skier visits (7–11 per cent lower than in the previous climatically average winter) and operating profits (33 per cent lower than during a climatically average winter) [27][28]. In the state of Washington (Pacific west region) skier visits declined 78 per cent in 2004/05 because of warm temperatures and frequent rain. The highly reduced season length caused some resorts to honour 2004/05 season passes again for the 2005/06 season in order to maintain customer loyalty [29]. Others were forced to forego investment plans, including repairs to a chair lift, indicating they would be in recovery mode for several years because of this one poor snow season [29].

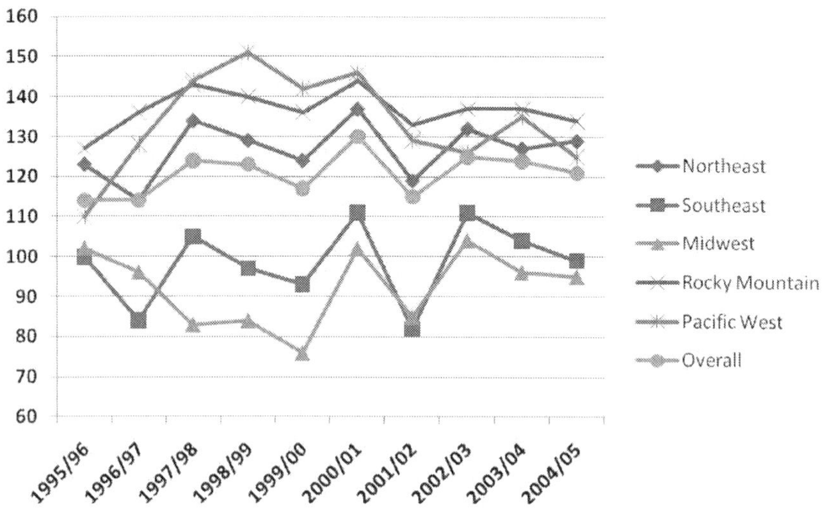

Figure 4. Annual ski season length (days open) by United States ski region (Source: National Ski Areas Association [30][31][32])

Figure 5. Marginal ski conditions during holiday season (Photo: Jackie Dawson)

The golf industry similarly attributes a considerable share of its annual economic success to climate. According to a number of golf industry reports, the single most important factor impacting season length and rounds played each year (both positively and negatively) is the weather. These same reports also identify climate as a primary determinant of irrigation needs and pest management, which represent major operating costs for most courses.

Weather and climate also play an important role in the planning, financial success and quality of visitor experience at special events that take place in outdoor venues, for example, music concerts, cultural festivals, sporting events – football, golf, auto racing [33]. Many outdoor events are held at certain times of the year in order to take advantage of certain climatic conditions or reduce climate risk. For example, the annual Albuquerque International Balloon Fiesta is held in October because of calm winds in the morning and limited severe weather. The Canadian Tulip Festival is held in May, coinciding with the maturate stage of the flowers. Climate conditions have forced organizers to undertake measures such as planting tulips bulbs in shady locations or irrigating flowers beds to delay bulb maturation to adapt the mismatch of tulip phenology and event schedule [34]. Climatic conditions can affect special events on a short-term basis (for example, short rain storm) or for the entire event (multiday rain or high winds, for example). A study of the reasons for the failure of special events through members of the International Festival and Events Association, found that weather was ranked first among eight external factors [35]. Of the full 30 factors considered in the study, weather was identified as the second most important factor contributing to event failure, only after a lack of corporate sponsorship.

"The golf industry's significant sensitivity to weather will make it one of the first to feel the impacts of increasing weather uncertainty due to climate change. As a result, we anticipate the industry to be impacted in the near-term as the old paradigm of business and financial planning around predictable and cyclical weather patterns may be disrupted by an increasingly volatile climate... certain geographic markets are already facing these challenges..."

WeatherBill Inc. (2007) [36]

Climate is also a salient influence on tourist decision-making and the travel experience. Climate is a key factor considered by tourists, consciously or implicitly during travel planning, and represents both a push and pull factor for tourists. Weather and climate are an intrinsic component of the vacation experience and have been found to be a central motivator for travel. General travel surveys conducted in Germany, the United Kingdom and Canada, have all found that weather and climate were a primary travel motivation for the majority of travellers [37][38][39]. Other traveller surveys, conducted in a number of countries have also revealed the importance of climate in the selection of a holiday destination and the timing of holiday travel [38][40][41][42][43]. Importantly, travel patterns are often related to the weather and climate conditions at the point of origin and not just the destination. For example, despite the global economic recession in 2008–2009 and expectation of reduced travel demand, the very rainy weather throughout much of the early summer in the United Kingdom was credited by the Association of British Travel Agents for the increase in foreign holiday bookings over the previous year [44].

Climate variability has been found to influence travel patterns (proportion of domestic and international holidays) and tourism expenditures in some nations. Studies of tourism demand in the United Kingdom have found that outbound and inbound visitor movements were responsive to weather (temperature and rainfall) during the current year as well as from the previous year [45][46][47][48]. Similar correlations between monthly accommodations demand (bed-nights) and summer temperatures (both current year and the previous summer) were found in Italy [49]. Demand for summertime inclusive tour charters by Norwegians, most of which (greater than 75 per cent) are to sunshine destinations, has been found to be influenced by weather conditions in the previous summer [50]. The influence of past seasons is understandable, for an individual that has experienced a negative impact on holiday satisfaction or perceived loss of holiday in the past due to weather, is likely to be wary of this potential when contemplating current holiday options. However, there is some evidence from tourist surveys that suggests the decision to return to a destination is largely unaffected by past experiences of poor weather [38][43]. A 1°C warmer than average summer season was found to increase domestic tourism expenditures in Canada by 4 per cent [51]. A number of sector or destination specific studies have also shown significant relationships between climatic conditions (daily to weekly timescales) and a range of tourism indicators – ski lift tickets, golf rounds, park attendance, special event attendance [52][53][54][55][56][57][58]. Importantly, climate information is embedded in a matrix with other information relevant to travel decision-making (cost, time, attractions, holiday commitment, motivations, for example) and these factors bound the use of climate information in travel decisions [59]. (See also Figure 1.)

There is also evidence that the weather conditions experienced at the destination have important influence on travel and holiday satisfaction (Figure 6). Visitor surveys by the Scottish Tourist Board show that 20 per cent of overseas visitors identify weather as the main cause of dissatisfaction [60]. Poor weather has similar impacts at sunshine destinations, where cool and windy conditions, such as those shown at a winter getaway resort in the Caribbean, have a highly negative impact on holiday satisfaction and are thought to adversely affect the likelihood of repeat visitation to the destination. The importance of snow conditions to skier satisfaction has been documented in several countries [27][61][62][63] and influences the satisfaction of visitors to winter resorts to an extent that it must be controlled for in surveying to avoid contamination of holiday satisfaction and destination evaluations [64]. Poor snow conditions have also been linked to negative impacts on personal safety of tourists. During the poor snow conditions of the 1990/1991 ski season in the Swiss and Austrian Alps, accident insurance claims by British skiers were almost double average levels, with approximately half listing accidents caused by exposed rocks and congestion on the slopes [60].

Figure 6. Reduced holiday satisfaction due to unseasonal cool and windy conditions (Photo: Daniel Scott)

The preceding lines of evidence demonstrate the climate sensitivity of the tourism sector. For the tourism industry and tourists alike, climate represents both a vital resource to be exploited and an important limiting factor that poses risks to be managed. Consequently, it is expected that the integrated effects of climate change (both shifts in climatic means and extremes) will have profound impacts on tourism businesses and destinations. Furthermore, because climate, the natural environment, personal safety and travel cost are primary factors in travel decisions, and each are projected to be significantly impacted by global climate change, far- reaching shifts in consumer travel demand may also occur. Accumulating evidence indicates that climate change, particularly high emission scenarios, could therefore fundamentally transform aspects of the global tourism sector in the decades ahead [27][65][66][67][68]. The demands for accurate and increasingly detailed climate information are therefore anticipated to increase substantially in order to allow tourism businesses and destinations to minimize associated risks and capitalize upon new opportunities posed by climate change, in an economically, socially and environmentally sustainable manner. WMO defines "climate information" as including: historical data, analyses and assessments based on these data, forecasts, predictions, outlooks, advisories, warnings, model outputs, model data, climate projections and scenarios, climate monitoring products, etc.

"Climate change will constitute an increasing risk for tourism operators in many destinations. With many tourism activities heavily dependent on the climate and insurance policies increasingly affected by natural hazards, accurate weather information and forecasting of extreme climatic events are becoming ever more important for tourism businesses."

UNWTO Secretary-General Francesco Frangialli

Beijing, November 2005

Despite the growing global economic importance of the tourism sector and the multiple, complex interactions between climate and tourism, there have been very limited evaluations of the use of climate information or assessments of the climate services needs within the sector [27][69][70][71]. This report provides a synopsis of the scientific and government literature and expert opinion (see list of experts consulted) on the capacities and needs for climate services in the tourism sector and was commissioned by the WMO and UNWTO to provide background information for the parallel working session on Climate Information for Adaptation and Risk Management in the Transportation and Tourism Sector at the 3rd World Climate Conference in Geneva, Switzerland.

The remainder of the report focuses on current and emerging applications of climate services in the tourism sector starting with an overview of climate information providers and various delivery pathways of climate information to the end-users within the tourism sector. The varied roles of different providers, types of existing collaborations and examples of best practice of communication and specialized climate products are identified. The next section summarizes the wide range of current and potential uses of climate information by the diverse tourism sector end-users, and key entry points into user decision-making. This summary is followed by a discussion of key knowledge gaps, research and capacity-building needs and partnerships that are required to accelerate the application of climate information to manage risks to climate variability and facilitate successful adaptation to climate change.

APPLICATIONS OF WEATHER AND CLIMATE INFORMATION IN THE TOURISM SECTOR

There has been a proliferation of climate information providers and communication technologies over the past two decades. The types of information available to tourism end-users from major providers vary considerably. While national meteorological services are primarily concerned with the production of climate information that benefits society at large in the public interest (for example, minimizing damage and enhancing human well-being), other government agencies and most private sector providers tailor their products specifically for use by special end-users and pay-for-service customers [72]. The relative availability of basic and specialized climate information, the types of providers and the delivery pathways (communication channels) vary substantially from nation to nation and even within nations. Figure 7 provides a conceptual framework of the supply of climate information and services to end-users in the tourism sector. WMO defines the following: "Basic services" are those services provided by a National Meteorological or Hydrometeorological Service in discharging its government's sovereign responsibilities to protect the life and property of its citizens, to contribute to their general welfare and quality of their environment, and to meet its international obligations under the Convention of the WMO and other relevant international treaties. "Specialized services" are those beyond basic service, which may include the provision of special data and products, their interpretation, distribution and dissemination and consultative advice

Sources of primary climate data typically include government agencies (NMSs, for example) and private meteorological companies (for example, The Weather Network®-Canada; The Weather Channel®-United States, United Kingdom, and Australia; BBC© Weather-United Kingdom), but can also include universities, non-governmental organizations (NGOs), and tourism operators that operate their own meteorological data collection station(s) (a ski operator, for example). Climate information is delivered to tourism end-users by many types of providers and communication media, ranging from tourism marketing and guide books, the Internet, television (including several all-weather channels in a number of developed countries), radio, newspapers and hand-held devices (BlackBerry® and iPhone™, for example). A brief overview of the objectives and capacities of each major type of climate information provider with respect to the tourism sector is provided in Section 2.1. Section 2.2 then provides a comprehensive, though not exhaustive, inventory of current and emerging applications of climate information in decision-making by major end-users within the tourism sector.

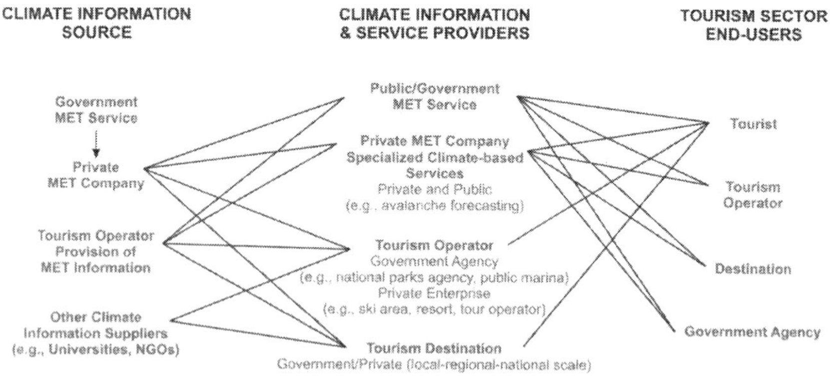

Figure 7. Conceptual framework of climate information in the tourism industry

CLIMATE INFORMATION AND SERVICE PROVIDERS

National Meteorological Services and other government agencies

Governments play an important role in the provision of climate information and this provision is sometimes dominated by legal requirements and constitutional provisions. Because basic weather and climate information is perceived as an economic public good, all 187 Member States and Territories of the World Meteorological Organization have established some form of national meteorological service to develop and operate the essential observation and data processing infrastructure needed to support the provision of climate services relevant to minimize the impact of climate-induced natural hazards (protect life and property of citizens), maintain the quality of environmental systems, enhance the economic performance of climate-sensitive sectors and meet other obligations under the Convention of the WMO and other international treaties. The NMSs may be organized in many different ways according to national circumstances and the major user communities they are required to serve, and have a high level of awareness within their respective countries [72].

For the tourism sector, NMS climate information can generally be grouped under four main categories: basic weather services (observations/nowcasting and short- to medium-range forecasts, for example); warnings to the public, aviation and marine transport weather services; specialized services for tourism end-users; and more recently climate change projections to facilitate successful adaptation. A detailed discussion of the provision of climate information for aviation and marine transport is provided in the Transport Sector paper in this proceedings.Like with all climate-sensitive economic sectors, the maintenance and enhancement of the global climate monitoring network that provides basic weather services are of vital interest to the tourism sector. Certain key tourism environments are underserviced by the global meteorological observation system (mountains and small islands) and would benefit from system improvements. The improved accuracy of basic weather forecasts is also important for tourism operators, as these forecasts inform operational decisions and because inaccurate forecasts can be detrimental to the tourist experience and tourism demand (see previous section). Improved accuracy (or skill) of forecasts (short- to medium-term and seasonal) has been identified at workshops in the United States, Spain, Jamaica, Bahamas, Fiji and Greece, by several of the experts consulted for this report and by Altalo and Hale [71] and Gamble and Leonard [73] as a requirement for increased use in operational decision-making in the tourism sector.

"Accurate, geographically specific meteorological information is essential for tourism operations. General forecasts, though meteorologically accurate, often have a negative impact on tourism operations because tourist destinations, such as beaches, coastal areas and mountains are often regions with unique and better than average microclimates. ... (climatic conditions) which would attract tourists, may differ substantially from prevailing regional conditions."

Final Communiqué Secure and Sustainable Living –

Social and Economic Benefits of Weather, Climate and Water Services

Madrid, 19-22 March 2007

Issuing weather watches, warnings, and advisories is a key function of NMSs. These enable appropriate action to be taken to prepare and protect travellers against the dangers of all forms of meteorological hazards (evacuation of people, deploying security personnel, reallocation of resources, or closing down operations, for example). Table 2 illustrates select countries and legislative responsibilities with respect to the issuance of meteorological warnings and advisories according to a recent World Meteorological Organization survey. While the survey revealed a number of NMSs to be directly responsible for the provision of meteorological information to users (Bahamas, Germany, New Zealand, for example), others are not legally responsible (Australia, parts of the Caribbean, for example) or the extent of legislative responsibilities is unknown.

In most nations in the Caribbean, the ministries of tourism work very closely with the NMS to monitor hurricane forecasts and are integral members of national emergency management agencies or task groups that coordinate hurricane preparedness. For example, in the Bahamas the Ministry of Tourism, along with other major government agencies and the International Committee of the Red Cross, is a member of the active hurricane committee that operates within the National Emergency Management Agency. The Ministry of Tourism has its own plan of operation, which includes contacting hotel operators directly by phone or e-mail as soon as a storm threat is forecast and providing hour-by-hour updates and evacuation plans as necessary.

Table 2. Select countries and legislated responsibilities for the issuance of meteorological information to the public

Country	*Legislated Responsibility for Issuing Meteorological Information*
Bahamas	Meteorological Service is the exclusive agent for issuing warnings on hurricanes, tropical storms and other weather conditions that are likely to give rise to floods, surges and wind damage.
China	The law provides exclusive powers to China Meteorological Association: "The state applies a unified system for the issuance of public meteorological forecasts and severe weather warnings. Meteorological stations subordinate to the competent meteorological departments at different levels shall, in compliance with their functions and duties, issue to the community, public meteorological forecasts and severe weather warnings ... No other organization or individual may issue to the community such forecasts and warning."
Germany	Under 'Duties and Competences', the Deutscher Wetterdienst is mandated with "issuance of official warnings of weather occurrences that could become a danger for public safety and order, especially concerning the impending danger of floods." No express provision for exclusive rights to the Met Service to issue forecasts and/or warnings.
New Zealand	The Minister of Transport who, under Meteorological Services Act 1990, is responsible for ensuring the provision of meteorological warnings and forecasts for New Zealand and the collection of data to support these services. Furthermore, the Minister is required to designate an organization that shall provide the authorized warning service in New Zealand.
South Africa	In Act No. 8 of 2001, under functions of the Weather Service, "Only the Weather Service may issue severe weather-related warnings over South Africa in order to ensure that there is single authoritative voice in this regard."
United Kingdom	The UK Met Office has no direct legislated responsibility. The Civil Contingencies Act (CCA) 2004 sets out the roles and responsibilities for UK Authorities who are required to respond to incidents (including weather-related incidents). A piece of secondary legislation (Statutory Instrument 2042/2005) states that respondents to the CCA should not seek to duplicate information provided by the Met Office.
United States	An Act of Congress provides that NOAA, as an Agency, shall not, except where specifically authorized by statute: establish an exclusive, restricted, or other distribution arrangement that interferes with timely and equitable availability of public information to the public; restrict or regulate the use, resale, or re-dissemination of public information by the public.
Australia	No express provision for exclusive rights to the Met Service to issue forecasts and/or warnings.
The Caribbean	No express provision for exclusive rights to the Met Service to issue forecasts and/or warnings.

Source: World Meteorological Organization, 2009 [75]

While the quantity and diversity of specialized climate services provided by NMSs to the tourism sector is currently limited, however, some of the joint government and WMO demonstration projects of nowcasting systems and applications have focuses on major sporting and tourism events (the Summer Olympic Games in Sydney, Australia, and Beijing, China). The NMSs have a critical role in supplying much of the climate information to the mass media and other tourism specific outlets (tourist guides, marketing brochures, travel planning Web sites, for example) and provide the essential historical, current, and forecast information that allow other providers to develop specialized climate products for the tourism sector. The climate change modelling and scenario communication capacity remain limited in some developing nations. The application of climate change projections for adaptation has also been found to be quite limited thus far within the tourism sector [68][74].

In addition to NMSs, in some nations, other government departments or agencies maintain basic or specialized climate observation systems that provide climate information relevant to tourism end-users. The Great Barrier Reef Marine Park Authority's (GBRMPA) Climate Change Response Programme is an exemplary case of partnerships to utilize climate data and other monitoring information for both management decision-making and communication to tourism stakeholders. The Great Barrier Reef has experienced several mass coral bleaching events in the past decade (1998, 2002, 2006), with 60 per cent of the reef being bleached during the largest event in 2002 [76]. The Response Plan has three main components: Early Warning System; Assessment and Monitoring; and Communication [77]. The Early Warning System uses climate forecasts, remote-sensing data, sea temperature monitoring, BleachWatch reports and site inspections to monitor conditions and provide early warnings of major bleaching events. Based on an emerging understanding of the relationship between weather and sea temperatures for the Great Barrier Reef, current and forecast weather conditions are used to indicate whether conditions conducive to bleaching are likely. The programme produces tools, such as ReefTemp (a Google™ Earth application), for the monitoring of environmental conditions conducive to coral bleaching or the early signs of bleaching. BleachWatch assists managers to collect data across a wide spatial distribution and is also used to communicate and increase the understanding of broader climate change impacts to the public, tourists and tourism operators. A survey of tourists in Cairns (North Queensland, Australia) asked if they would visit the region if they knew that there had been a recent bleaching event – 29 per cent were uncertain and 35 per cent indicated they would not.

PRIVATE SECTOR CLIMATE SERVICE PROVIDERS

Private sector climate service providers have, to a great extent, embraced the rapid technological advancements that have occurred over the past decade with respect to meteorological observation, remote sensing and model development, as well as in communication and delivery systems (advances in cable television, the Internet, and more recently wireless hand-held devices such as RIM's® BlackBerry® and Apple's® iPhone™), in order to provide more user-friendly access and formats of climate information than provided by NMSs and other government agencies, as well as to develop specialized, value-added climate information products for the tourism sector. As a result, private sector climate service providers tend to provide a greater variety of climate information and applications that are of interest to tourism end-users.

Private sector climate service providers have led in terms of innovation of specialized climate services tailored to specific tourism destinations, individual tourist activities and subsectors. The major private climate service provider in the United States (The Weather Channel®) exemplifies best practice in this area of innovation. It should be noted that other private meteorological companies in the United States (such as *Accuweather*) and in Europe (such as *WeatherNet*) also offer a range of forecast products tailored to recreation and tourism sector operations (skiing/snowmaking, golf turf management, marine forecasts, pool management, incident reports, business revenue/visitor analyses, weather risk insurance assessments) and traveler decision-making. The Weather Channel® provides access to specialized weather reports for the following tourism activities, events and destinations: golfing, ski and snowboarding, snowmobiling, special events and major sporting events (Major League Baseball and Professional Golf Association events, for example. The Weather Channel® is the "official supplier" of weather information to Major League Baseball and the Professional Golf Association.), weather sensitive travel routes (airport weather and highway construction, for example), coastal beach destinations, international sunshine destinations and potential wedding and honeymoon locations. In some situations, what private meteorological companies identify as activity/destination specific forecasts, are no more than renamed local forecasts for nearby cities and provide little or no added value for tourist decision-making. The Weather Channel® has also developed several climate indices relevant to the individual tourist and to various tourism subsectors. The Spectator Index (Figure 8) provides tourists a rating of how comfortable they will feel while watching a sporting event. A variety of weather factors are used to calculate the index, including temperature, probability of precipitation, humidity, wind speed and cloud cover. A Spectator Index value of 1 signifies poor conditions while a 10 signifies ideal conditions for watching an event. The Weather Channel® states that the best index values result from short-sleeve temperatures, minimal chances of precipitation, low humidity, light winds and fair skies. Conversely, cold and stormy conditions will result in the lower values.

Figure 8. The Weather Channel® Spectator Index (Source: The Weather Channel® *[78])*

Other unique multivariable indices developed by the Weather Channel®, include the Ski Index (Figure 9), Golf Index (Figure 10) and the Fishing Index (Figure 10). While the weather parameters used to calculate the ski index are not entirely known, a number of parameters are acknowledged specifically as being used to calculate the rating: surface snow conditions, average snow depth, new snow accumulation, temperature, precipitation and wind speed. The Ski Index is based on a 1 to 10 scale, where a 10 signifies ideal ski conditions and 1 signifies dangerous conditions. Unpleasant weather conditions, such as thunderstorms, wind speeds at or greater than 35 mph and freezing rain will automatically set the Ski Index value to 1. A Ski Index will not be calculated for temperatures greater than 60 degrees Fahrenheit.

To ensure the upmost accuracy, the Ski Index is only available for the current day and next two days and is updated frequently. The forecast Ski Index is not calculated after 4 p.m. since the lowest temperatures tend to occur at night and would therefore unfavourably skew results. The weather parameters used to calculate the Golf Index are also not known; however, parameters known to be included in the rating are: extreme temperatures, high dew points, low visibilities, thunderstorm risk, high winds and precipitation. The Golf Index is based on a 0 to 10 scale, where 0 is an unfavourable golfing day, and 10 is an excellent golfing day. The Fishing Index differs in that it only includes three rating classifications (excellent, good, fair). Variables used to calculate this index include time of day, solar influence, lunar influence, weather and wildlife charts and declination and diurnal inequality.

The Weather Channel® also provides users with current and forecast weather conditions for most national and state parks in the United States. A unique application developed by the Weather Channel® for its parks reports is the Mosquito Activity Forecast, which predicts how active mosquitoes in a particular park will be based on a variety of environmental factors, including temperature, wind, humidity values and time of day and year. The levels of mosquito activity ranges from none (conditions not favourable for mosquitoes) to very high (light winds and ideal temperatures make it perfect for mosquitoes). The Mosquito Activity Forecast cannot predict how many mosquitoes will be in a particular area (a campground, for example) or bite frequency. Mosquito populations and concentrations of species with different behaviours can vary considerably in short distances and are not taken into account in the forecast.

Figure 9. The Weather Channel® Ski Index (Source: The Weather Channel® *[79]*)

Figure 10. The Weather Channel® Golf Index and Fishing Indices (Source: The Weather Channel® *[80][81]*)

The development of such indices and user-friendly tools, such as the ability to conduct climate comparisons between two potential tourism destinations, has done much to increase overall brand awareness of private climate service providers in some nations. Moreover, because weather and climate data provided by private Met services can be accessed more readily via a number of different media (Internet, TV, desktop applications, mobile phones, RSS feeds and wireless handheld devices such as SmartPhones, for example), end-users are constantly reminded of who is providing them with weather and climate-related information and the types of services offered.

Table 3. National coverage of mountain/snow forecasts by SnowForecast.com

Andorra	9	Germany	80	Pakistan	4
Argentina	12	Greece	14	Peru	1
Armenia	2	Greenland	2	Poland	11
Australia	17	Hungary	4	Portugal	1
Austria	215	Iceland	4	The former Yugoslav Republic of Macedonia	6
Plurinational State of Bolivia	1	India	11	Réunion	1
Bosnia and Herzegovina	6	Indonesia	1	Romania	30
Bulgaria	6	Islamic Republic of Iran	8	Russian Federation	11
Canada	108	Israel	1	South Africa	2
Chile	17	Italy	193	Serbia/Montenegro	5
China	6	Japan	311	Slovakia	35
Colombia	1	Kazakhstan	1	Slovenia	25
Croatia	4	Kenya	1	Spain	30
Cyprus	1	Lebanon	4	Sweden	31
Czech Republic	37	Lesotho	1	Switzerland	204
Ecuador	1	Liechtenstein	2	Taiwan, Province of China	1
Egypt	1	Mexico - Puebla	1	United Republic of Tanzania	1
Ethiopia	1	Morocco	2	Turkey	15
Finland	5	Myanmar	1	Uganda	1
France	174	Nepal	2	Ukraine	12
France - Corsica	1	New Zealand	40	United Kingdom	16
Georgia	1	Norway	37	United States	408

Other private sector climate service providers concentrate on specialized climate information to tourists within a single tourism subsector. For example, SnowForecast.com provides climate information for over 2 200 ski locations around the world. (See Table 3 for coverage by nation.) It applies topographical corrections to basic NMS data to produce "more accurate" snow forecasts for mountain resort destinations, including snow conditions and forecasts for top-, mid-, and bottom-lifts (Figure 11). Users can access historical weather data, current weather conditions (for up to five weather stations located closest to the resort) and six-day weather forecasts (including morning, midday, and night forecasts for each day) free-of-charge. SnowForecast.com also provides access to more detailed two-day and nine-day forecasts for a subscription fee. In addition to static maps, SnowForecast.com provides access to interactive maps that allow users to select from a number of climatic variables (temperature, snowfall and wind, for example) and animate projected changes for up to a six-day period for a given region. The Website also provides other tourism relevant information to end-users, including ski trail maps, resort reviews, and links to current conditions Webcams that could be used in the destination and activity choice decision-making process. The Website also provides visitors with the opportunity to register for Snow Alerts that are delivered regularly via e-mail. Such information and functionality is virtually never provided by government agencies and rarely provided by even the largest and most prominent private weather service providers.

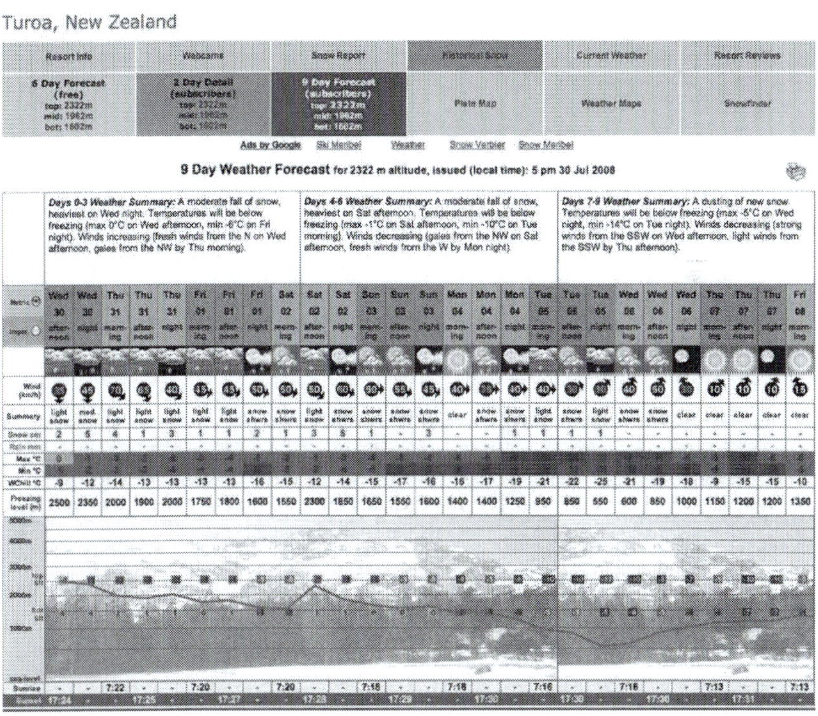

Figure 11. SnowForecast.com weather forecast for Turoa ski resort, New Zealand (Source: SnowForecast.com [82])

Table 4. National coverage of surf forecasts by Surf-Forecast.com

American Samoa	19	Ghana	17	Papua New Guinea	18
Angola	7	Guadeloupe	24	Peru	82
Argentina	17	Guam	13	Philippines	45
Australia	623	Guatemala	15	Portugal	124
Bahamas	16	India	8	Puerto Rico	16
Barbados	33	Indonesia	146	Réunion	1
Belgium	7	Ireland	87	Réunion	25
Bermuda	9	Israel	6	Samoa	16
Brazil	377	Italy	40	Senegal	11
British Virgin Islands	19	Japan	113	Seychelles	11
Canada	48	Madagascar	18	Solomon Islands	9
Cape Verde	10	Maldives	23	South Africa	256
Chile	71	Martinique	14	Spain	222
China	9	Mauritius	15	Spain	77
Colombia	9	Mexico	234	Sri Lanka	23
Cook Islands	7	Montserrat	16	Sweden	13
Costa Rica	70	Morocco	55	Taiwan, Province of China	12
Côte d'Ivoire	8	Mozambique	8	Tonga	7
Denmark	8	Namibia	12	Trinidad and Tobago	14
Dominican Republic	44	Netherlands	20	United Kingdom	436
Ecuador	54	Netherlands Antilles	7	United States	898
El Salvador	16	New Caledonia	4	Uruguay	52
Fiji	22	New Zealand	356	Vanuatu	31
France	197	Nicaragua	27	Bolivarian Republic of Venezuela	33
French Polynesia	40	Oman	10	Viet Nam	7
Gabon	10	Pakistan	10		
Germany	15	Panama	37		

A similar product has been developed by Surf-Forecast.com for surfers, kite surfers, sea kayakers and other water sport enthusiasts. The NMS weather data are used to develop coastal weather and swell projections for over 8 000 surfing locations around the globe. (See Table 4 for coverage by nation.) Surf-Forecast.com uses a star rating scale to rate the quality of surf conditions at a particular location. The scale is based on swell size and character (the bigger the swell and longer the period, the higher the rating). However, if the wind is onshore, the star rating drops in proportion to the wind speed (Figure 12). A rating of 10 is considered to be the best surf with classic conditions and light offshore winds. Flat conditions, blown out waves, onshore winds or very strong winds in any direction result in a star rating of 0. Other additional tourism information that is provided by Surf-Forecast.com includes swell and wind maps, surf cams, tide tables and surf-site-specific user ratings.

Avalanches are of high importance to winter sports tourism, as many popular destinations, such as renowned ski resorts, parks and climbing areas in the European Alps, Rocky Mountains and Himalayas are in avalanche-prone locations. Approximately 150 avalanche fatalities occur each year within the 17 countries that are members of the International Commission for Alpine Rescue [84] and the majority of victims are skiers, snowboarders and winter climbers [84][85]. The mandate for avalanche prediction and communication varies by nation.

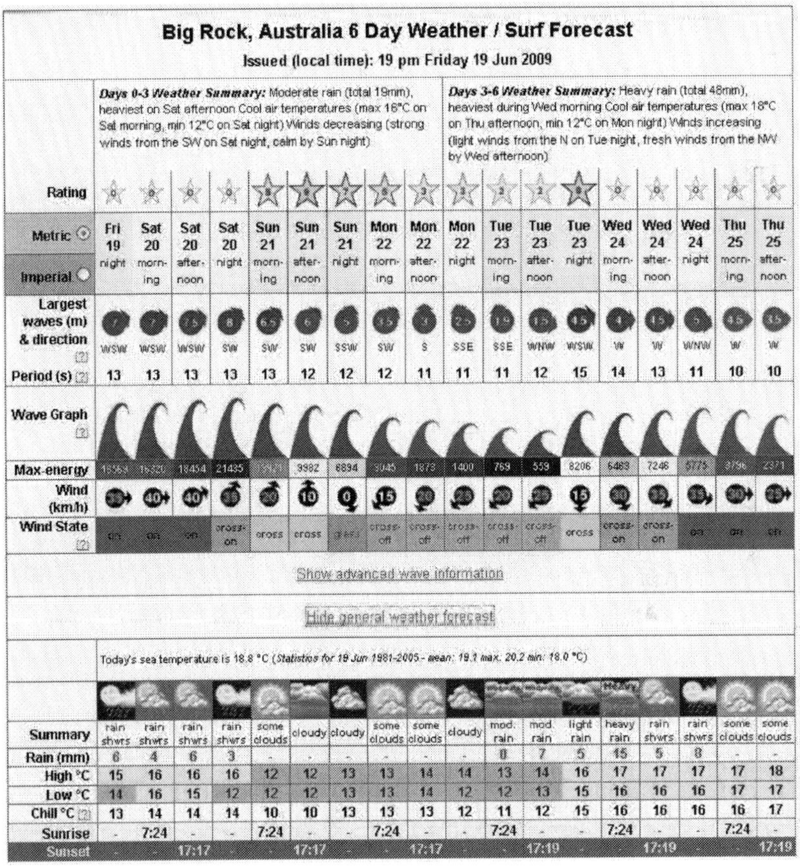

Figure 12. Surf-Forecast.com surf forecast for Big Rock, Australia (Source: Surf-Forecast.com [83])

In Canada, a broad coalition of federal, provincial, private and not-for-profit sector stakeholders established the Canadian Avalanche Centre (CAC) as a not-for-profit corporation to provide public avalanche safety warnings, coordinate public avalanche safety programming and provide professional snow science and avalanche training. The CAC uses a combination of available weather and climate information, technology and science to develop their avalanche forecasts (Figure 13). Weather and climate information is primarily provided by Environment Canada or directly by local agencies or tourism operators, and satellite and remote-sensing data are provided by the University of Washington.

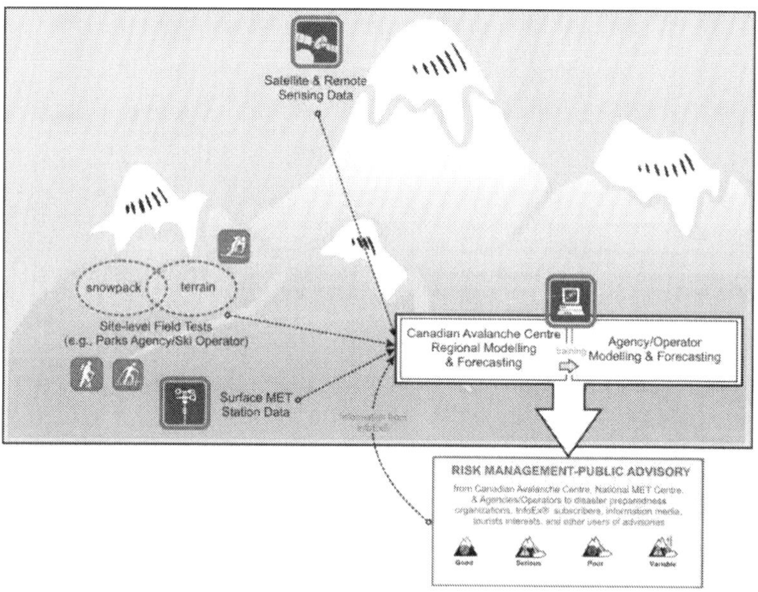

Figure 13. The use of weather and climate information in Canadian avalanche public advisories and the risk management communication chain

The CAC produces a daily Backcountry Avalanche Advisory, Public Avalanche Forecasts three times a week and weekly Public Avalanche Information Reports for very specific mountain regions and specific destinations (national parks and major ski resorts, for example). Advisories are posted on the CAC Website, the Canadian Meteorological Service Website (Environment Canada), agency and operator Websites, and are forwarded to the information media and other users of advisories (Figure 14). All areas are rated on the avalanche danger scale, which provides travel and activity recommendations for users. The Canadian Avalanche Association (CAA), a subsidiary of the CAC, also offers subscription-based mountain weather forecast services for tourism organizations and individuals who are involved in avalanche-related work via its Industry Information Exchange (InfoEx®) program. The more than 450 organizations that currently subscribe to InfoEx® are all Professional Members and actively manage avalanche hazards – backcountry and heli-skiing companies, mountain parks, resorts and lodges and road maintenance departments, for example [86]. The fee for service options include specialized forecasts for a distinct geographic location and one-on-one consultation services.

	Danger level	Snowpack stability	Avalanche triggering probability	Consequences for transportation routes and settlements / recommendations	Consequences for persons outside secured zones / recommendations
1		The snowpack is generally well bonded and stable.	Triggering is generally possible only with high additional loads** on very few extreme slopes. Only natural sluffs and small avalanches are possible.	No danger	Generally safe conditions
2		The snowpack is only moderately well bonded on some steep slopes*, otherwise it is generally well bonded.	Triggering is possible, particularly through high additional loads**, mainly on steep slopes indicated in the bulletin. Large natural avalanches are not expected.	Low danger of natural avalanches	Mostly favourable conditions. Careful route selection, especially on steep slopes of indicated aspects and altitude zones.
3		The snowpack is moderately to weakly bonded on many steep slopes*.	Triggering is possible, even through low additional loads** mainly on steep slopes indicated in the bulletin. In certain conditions, some medium and occasionally large natural avalanches are possible.	Isolated exposed sectors are endangered. Some safety measures recommended in those places.	Partially unfavourable conditions. Experience in the assessment of avalanche danger is required. Steep slopes of indicated aspects and altitude zones should be avoided if possible.
4		The snowpack is weakly bonded on most steep slopes*.	Triggering is probable even through low additional loads** on many steep slopes. In certain conditions, many medium and multiple large natural avalanches are expected.	Many exposed sectors are endangered. Safety measures recommended in those places.	Unfavourable conditions. Extensive experience in the assessment of avalanche danger is required. Remain in moderately steep terrain / heed avalanche run out zones.
5		The snowpack is generally weakly bonded and largely unstable.	Many large natural avalanches are expected, even in moderately steep terrain.	Acute danger. Comprehensive safety measures.	Highly unfavourable conditions. Avoid open terrain.

Figure 14. European avalanche danger rating scale for recreation and travel (Source: Swiss Federal Institute for Forest, Snow and Landscape Research [87])

In Europe, the WSL Institute for Snow and Avalanche Research SLF (Switzerland) has been publishing public Avalanche Bulletins since the 1940s and the standardized Avalanche Danger Scale shown in Figure 15 has been in use throughout Europe since 1993. National and Regional Avalanche Bulletins are broadcast to the public on several media (television, radio, newspapers, Websites) as well as through mobile communications (SMS, MMS).

The Weather Network (Canada) application for the RIM® BlackBerry® Bold™ and Storm™

Oakley® The Snow Report iPod Touch™ and iPhone™ application.

The North Face® Snow Report iPod Touch™ and iPhone™ application.

Fishing Calendar iPod Touch™ and iPhone™ application.

Figure 15. Select examples of RIM® BlackBerry® and Apple® iPhone™ and iPod Touch™ weather-related applications available to tourism and recreation end-users

Private sector climate service providers have been innovators in the use of emerging communications technologies to deliver specialized climate information to tourists and other tourism sector end-users. In the last two years, there has been a tremendous increase in the use of mobile phones to search the Internet, and a proliferation in the development of climate-related applications made available to tourism end-users through wireless hand-held devices. Apple's® iPod Touch™ and iPhone™ and Research in Motion's (RIM®) BlackBerry® products offer a diverse range of both free and pay weather-related information applications, with Apple's® App Store™ alone providing 245 weather-related products to consumers. The BlackBerry® App World™ also provides end-users with the opportunity to personalize their BlackBerry® Smartphone with several weather-related applications. With the number of mobile phones expected to exceed 4 billion in 2010, these specialized applications are likely to further revolutionize how travellers receive and utilize weather and climate information over the next decade.

The Weather Channel® 2.0 (United States) for the iPhone™, iPod Touch™, BlackBerry® Bold™ and Storm™, for example, provides users with real-time (current) conditions, hourly and 10-day forecasts, severe weather warnings and other fully interactive applications, including in-motion radar maps and traffic cameras (for select markets). The original free, ad-based version of The Weather Channel® application for iPhone™ and iPod Touch™ has received more than 3.8 million downloads (as of 26 May 2009) since its November 2008 launch [88], and is the second most downloaded weather-related application available in the Apple® App Store, illustrating the demand for climate information from the general public. For a small fee (US$ 3.99), consumers can also download a premium version of software. The Weather Channel® Max provides ad-free and more specialized weather information, such as in-motion future radar imagery projected on a map. The availability of radar information and warnings on Smartphones represent a powerful new information source for tourist activity planning and safety. Where Smartphone communications exist, tourists can be informed of adverse or hazardous weather and revise activities or avoid imminent danger. Other features of the premium application include additional severe weather coverage, an enhanced video centre, tropical updates and a Boat and Beach Forecast that provides marine coastal forecasts, local tide information and wind and surf conditions for coastal locations around the globe.

Both the Apple® App Store™ and the BlackBerry® App World™ provide a number of more specialized weather applications tailored to specific tourism-related activities. The Apple® App Store™ contains applications developed specifically for alpine skiing (iSki App, The North Face® Snow Report, SkiResort and SnoCountry, for example), boating (TideApp, for example), sailing, windsurfing and kitesurfing (iWind Italia Light, for example), surfing (Oakely® Surf Report, for example) and fishing (Fishing Calendar, for example). (See Figure 15 for screen captures.) The North Face® SnowReport is a free iPhone™ application that provides skiers and snowboarders with up-to-date weather and map information about winter resorts across the globe. The application provides resort-based snow conditions with current and expected snowfall, detailed daily forecasts and wind direction and speed information. Users can access their favourite resorts and navigate around them once they arrive through Google™ Maps and can also utilize the Trail Maps function to view a detailed lay of the land through high-resolution images. Additional information through links to the National Oceanic and Atmospheric Administration (NOAA) Climate Prediction Center, as well as resort-specific Websites, can be accessed via the application.

Overall, while the number of weather- and tourism-related applications available to users is burgeoning, there is still a great opportunity for the wireless hand-held device and Smartphone programming communities to further develop and tailor specific weather and climate-related applications for tourism and recreation end-users.

TOURISM OPERATORS

Reviews of tourism marketing materials and Websites have indicated that most tourism operators provide limited climate information to potential travellers; the most common practice is to provide only average monthly temperatures [60][89][90][91]. De Freitas [92] argues that such averages have limited value for tourists and that the tourism community should instead focus on communicating the probability of experiencing certain conditions known to be important to travellers (for example, temperatures above a certain threshold, sunshine) during specific periods (monthly or bi-weekly). Some tourism operators, especially those in more dynamic tourism environments where weather conditions can pose a health risk to tourists (for example, mountains, deserts, marine-coastal zones), provide more comprehensive climate information to their potential visitors so that they can properly prepare for their travels, including the best time climatically for travel and what to pack to ensure a safe and comfortable travel experience. Resorts and other tourism operators also frequently provide climate information to their guests to assist them with planning holiday activities once they have arrived, and as a marketing tool to reinforce the value of the holiday (Figure 16).

Figure 16. Tourism operator on-site marketing with climate information (Photo: Daniel Scott)

In most cases, tourism operators obtain primary climate information from NMSs or private meteorological companies and relay this information to tourists who are planning to utilize their travel services or who are already travelling. An example of a creative new private sector partnership began in 2009 between Husa Hotels and Cham Palaces and Hotels (Jordan) and Weather Decision Technologies to provide weather content for each hotel in the two chains (including more than 125 cities) as a value-added service for their customers. Some tourism operators provide multiple sources of climate information to travellers and a limited number also maintain climate monitoring stations and data processing capabilities at the site level. For example, Whistler-Blackcomb ski resort in Canada provides multiple types of climate information (Figure 17) and has implemented its own climate monitoring network in order to communicate more accurate local conditions to end-users and to utilize real-time information in operations decision-making (as input into its snow production management system, for example). More detailed climate information, including conditions at different altitudes (base, middle and top if there are chair lifts on each mountain) are provided to more accurately communicate the diverse climatic conditions that can be experienced by visitors and to provide greater accuracy of current conditions than NMSs and private meteorological services. Although such detailed climate information may not be needed by general tourists, it is desired by activity specialists (expert skiers in this case) who have more advanced climate information needs.

Destinations

Like tourism operators, organizations that promote tourism to specific destinations typically provide climate information with two purposes: to market the destination, and to assist travellers to prepare for safe and comfortable travel experiences. The studies that have examined the way in which climate is used in the advertising of destinations have consistently found that weather and climate are used in a high percentage of brochures [90][91][93][94][95][96]. According to Besancenot (1991), "The iconographic analysis of tourist brochures and the careful reading of the accompanying text only confirm the obsessive presence of references, direct or indirect, to the climate."

Figure 17. Sources of climate information provided for Whistler-Blackcomb (Canada) ski resort

The sophistication of climate information within destination marketing communications (brochures, TV, Websites, for example) ranges widely and depends on the prominence of climate within the brand image of the destination. At one end of the spectrum are marketing communications that provide absolutely no climate information on the destination, but display a very obvious blue sky bias in all photography of the destination [90][91][97]. Often, even basic climate information (monthly mean temperature and precipitation) and advice on what to wear and health-related issues (insect repellent, UV index and sunscreen, for example) are not provided. More detailed climate information on sunshine hours, water temperatures, elevational differences in temperature and climate-related hazards remains the exception. At the other end of the spectrum are destinations that provide detailed climate information that is tailored to the types of tourists who visit and to the activities they typically undertake.

In destinations where seasonal weather is not highly conducive to tourism, climate is sometimes downplayed in marketing communications (Iceland's "more solar than polar", for example) or is omitted altogether. Marketing is also used to specifically address unfavourable perceptions about a destination's climate. For example, a tourism brochure for Brittany, France, informs tourists that "Common misconceptions and prejudices have portrayed Brittany as a rainy region when in fact its maritime climate is mild and bracing ... [and its] iodine-rich sea air is unique and just to breathe it in is to enjoy its health-giving properties." In contrast, other destinations have cultivated a destination image and marketing strategy around certain activities or experiences that are largely based on the local climate. The Cayman Islands promotes its "perpetual summer", while Florida brands itself as "the sunshine state" (Figure 18). In an effort to differentiate its climate-based destination image, the Barbados Tourism Authority began offering its perfect weather guarantee in 2009. (See Table 3.) Similar money-back sunshine guarantees have begun to be offered for destinations in the south of France by travel agents in France in cooperation with the insurance company Aon France [98].

Other destinations use what would typically be considered adverse weather conditions to develop distinct tourism products. Historically, tourism has not been an important activity to the local economy of Tarifa, Spain, as the frequent high winds were not suitable for conventional sun, sand, sea tourism development. However, the high winds of Tarifa are a very valuable resource for water sports such as windsurfing and kitesurfing, and the community now markets itself as the "windsurfing capital of Europe" or the "Costa de la Luz y el Viento" ("the Coast of Light and Wind"). Vancouver Island (British Columbia, Canada) promotes a

Storm Watching product on its Pacific Ocean coast during the fall and winter months and encourages visitors to take advantage of the special conditions created during La Niña phase of Pacific Ocean Southern Oscillation.

"...experience the raw power of the mighty Pacific Ocean as ferocious waves roll in from Japan and pound the shores of the rugged west coast - nature in all its fierce majesty! ... Watch as a smooth sandy beach is transformed when hundreds of pieces of driftwood are pitched up onto its sandy shore. Likewise, the Gulf Islands experience some stunning electrical storms that can only be viewed out over the water... When the storm has passed, you'll be rewarded with peaceful strolls along cool, deserted beaches – a good time for beachcombing for washed-up treasures. Make the best of these *La Nina* seasons by creating opportunities to witness some spectacular weather."

Storm Watching Accommodation and Tours

(Source: BritishColumbia.com)

Climate service users

The potential use of climate information within tourism is tremendous given the diversity of end-users in the sector. The full temporal scale of climate information, from nowcasting (up to 1 hour), to short- and medium-range forecasts (1 and 7–10 days), to multi-decadal climate change projections is utilized within the tourism sector and examples of applications are outlined for each major end-user below. However, to date, there has been no systematic assessment of the extent of climate information use in tourism decision-making in any region or subsector or of whether the information needs of specific tourism user groups are being met regionally or nationally. User awareness and capacity to use climate products and services also remain largely unknown. All remain important areas for future research and capacity-building.

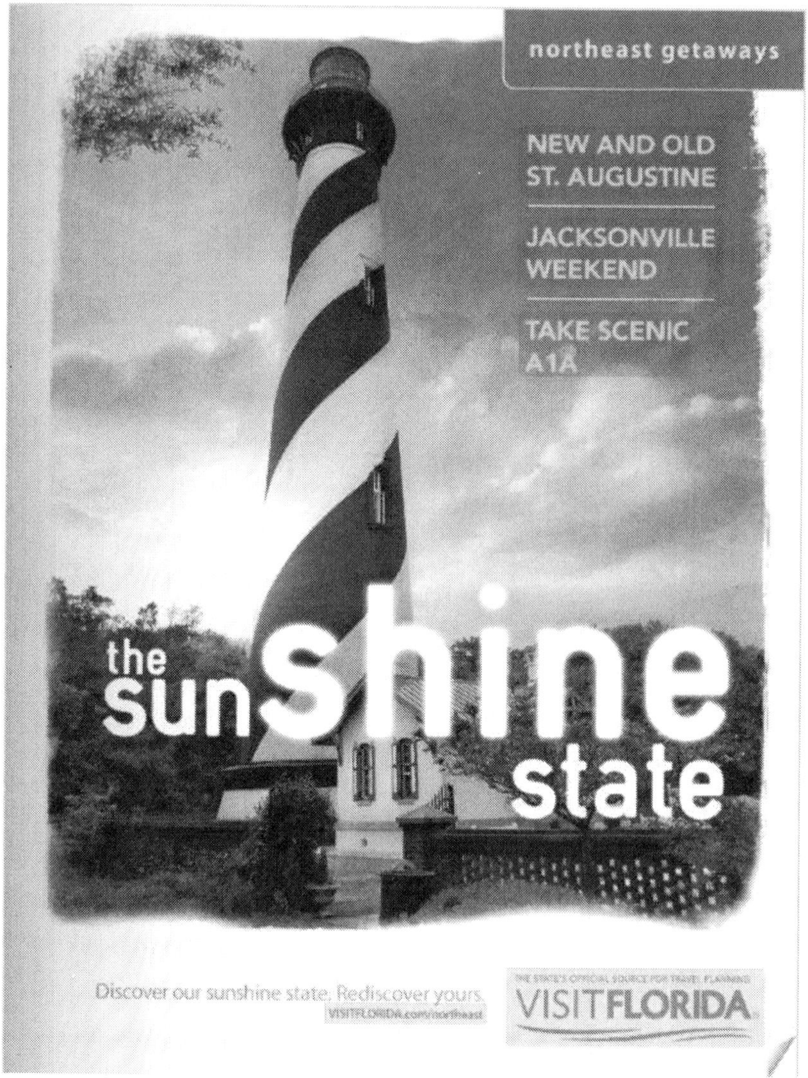

Figure 18. "The Sunshine State" tourism marketing for north-east Florida (Source: VisitFlorida.com)

Tourists

As indicated, weather and climate have broad significance to tourist decision-making and the vacation experience. Climate is of universal importance in defining destination attractiveness and a central motivator in the selection of a holiday destination and the timing of holiday travel. Climate influences travel patterns (proportion of domestic and international holidays), tourism expenditures and overall holiday satisfaction. Travellers are interested in the climate and weather at their intended destination as well as the weather along the way (travel phase). Business travellers are particularly cognizant of how weather causes delays and diversions, and utilize forecasts in routing decisions. Figure 19 conceptualizes the influence of different types of climate information (historic, forecasts, nowcasts) on tourist decision-making.

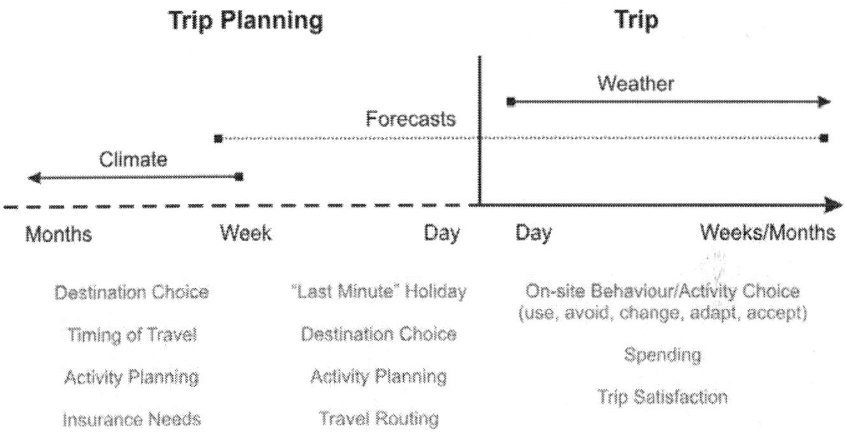

Figure 19. Weather-climate information for leisure tourist decision-making (Adapted from Scott [99])

The limited studies that have examined the use of climate information by travellers reveal widespread use in holiday planning. Tourism and recreation users were shown to generate the largest demand from automated telephone weather services in Scotland and Britain [100]. A survey of German outbound tourists found 73 per cent informed themselves about the climate of their destination and 42 per cent informed themselves about the climate before booking their travel [41]. A similar survey of Northern European travellers to the Mediterranean region found 86 per cent would obtain information on their destination's climate, with 81 per cent doing so before making any travel reservations [101].

Interestingly, consideration of current weather conditions or near-term forecasts (next 1–4 days) has been found to be the most important factor in last minute domestic leisure tourism [102]. With the trend toward shorter timeframes for travel planning, especially discounted last minute bookings made in the week (or day) prior to departure, the value of short- and medium-term forecasts for travel planning is likely to increase. In the same way, the influence of media stories about unfavourable weather for tourism and extreme events may also increase. For example, when northern European travellers were asked if media stories about heatwaves in the Mediterranean would affect their travel plans to a destination in the region, 51 per cent said they would alter their travel plans in some way and a further 15 per cent said they would seek additional information before deciding [101]. This finding was consistent with earlier research that found travellers interpreted weather forecasts differently, depending on the relative level of commitment to holiday travel [59].

Tourist decision-making is a highly complex process, incorporating multiple influencing factors. The place of weather and climate information in the psychological process of travel planning remains an important area for further research.

Tourism developers, operators and destinations

Weather and climate information provide input into a several decision-making contexts for tourism developers, operators and destinations (Figure 20). Historic climate information can be used for strategic planning of tourism infrastructure, including location analysis for new resorts, architectural and landscape design and construction scheduling in remote locations. Altalo and Hale [71] found that weather and climate were not usually cited as reasons for development (or not) of new resorts and accommodations, nor critically to influence site selection decisions relative to other macrolevel factors like transportation access, source markets, land ownership and coastal access. Climate information was utilized more extensively in engineering, construction planning, property design and maintenance and other post-build decisions (insurance, heating–cooling budgeting, staffing). The limited availability of historic climate information in many developing nations and remote locations (for example, smaller islands and mountainous areas) at the scale relevant to tourism developers has been a significant barrier in the past [103][104] and the limited availability of electronically archived historic climate data from many stations continues to be a barrier in regions such as the Caribbean.

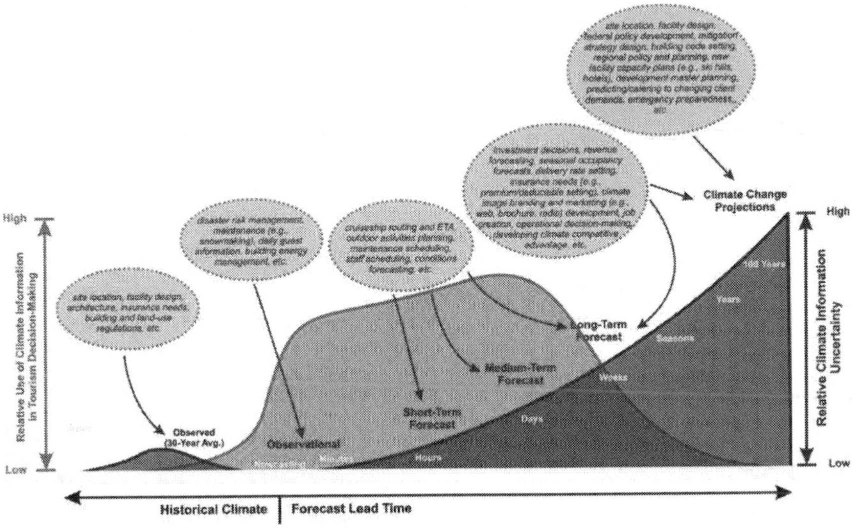

Figure 20. Potential uses of weather and climate information by tourism operators and destinations

In the same way that historic climate information has been used for strategic planning of future tourism developments, climate change projections are now also being utilized to anticipate and adapt to market risks and opportunities at the business, destination and national level. Since climate change is likely to increase the occurrence of extreme weather events [105], there is a need for a long-term integrated response. Real estate and tourism developers have begun a billion dollar investment in a giant new ski area near the remote mountain town of Revelstoke (British Columbia, Canada), because they believe the astounding snowfall in the area (an average of 15 meters annually) will become a major market advantage as snow conditions at major ski areas in more southern locations of the United States and the European Alps are projected to continue to decline in the decades ahead [106]. At the destination scale, both the internationally renowned mountain tourism community of Aspen, Colorado, [107] and Tourism Australia are using climate change scenarios in the development of climate change adaptation plans. While the use of climate change information by tourism developers, operators and destinations remains isolated to date [68][74], its necessity for successful climate change adaptation is anticipated to increase substantially in the decades ahead (Figure 20).

Historic climate data are also the foundation for the growing application of innovative weather derivatives and index insurance products to reduce weather risk in the tourism sector. Weather derivatives are financial instruments that emerged in the late 1990s to protect against weather-related loss of revenue in the energy and agriculture sectors and differ fundamentally from weather insurance commonly held by tourism operators [108]. Weather index insurance is a financial risk transfer product that operates on similar principles. Insurance policies against infrastructure damage or business interruption from wind, floods or other extreme events pay out only in the event of an actual economic loss and reimburse only the actual amount of loss via established claim procedures. In contrast, weather derivative contracts and weather index insurance pay a fixed amount based on observed weather conditions regardless of the extent of losses experienced or whether actual damages can be demonstrated.

One of the key advantages of weather derivatives and weather index insurance is that the scalability of the product can range from individual businesses to national governments (Table 5). This feature is particularly suited to small and medium sized enterprises that dominate the tourism sector. The flexibility of designing contracts and insurance based on weather at any climate station is also uniquely suited to the tourism sector, where impacts in revenues can be attributed to weather conditions in source markets that are hundreds or thousands of kilometres from the destination where the impacts occur. Since their inception in the late 1990s, the number of companies that offer weather derivatives and the countries in which they are offered has grown substantially. In 2007–2008, the total market value of weather derivatives was estimated at US$ 32 billion [109].

Participation of the tourism sector in the weather derivatives market has remained rather limited. Knowledge of just how dependent different tourism subsectors and specific businesses are on the weather is still relatively unexplored and is hampering development within the tourism sector. Nonetheless, as the examples of tourism applications of weather derivatives in Table 5 illustrate, there is tremendous potential to develop innovative partnerships with the financial services sector to develop highly customized contracts to reduce weather-related revenue loss and create new destination marketing strategies that deliver a competitive advantage regardless of actual weather conditions. For example, the company WeatherBill now provides specialized consultations for ski resorts, sports venues (golf, water and theme parks), events (sports or entertainment) and for the travel sector in order to stabilize income, control costs, improve marketing and public relations and protect special events.

Climate information, particularly real-time observations and short-term forecasts, are used extensively by tourism operators. Recent weather observations are important inputs into a range of decision-support tools, including: automated turf management systems used by golf courses and other sports facilities, snow production systems used by ski areas, cruise ship and aviation routing, fire and avalanche rating and warning systems and energy management (heating–cooling) systems used by accommodations providers. A study examining the enhancement of swimming pool management decisions with forecast information found the potential for large annual savings related to staffing. The study also recommended that the pool management company consider using value-added, short-term weather forecasts, produced by private sector meteorologists, in their decisions [112]. However, as Altalo and Hale [71] point out, despite the acceptance of seasonal and weather influences on tourist flows by the accommodation and hospitality subsectors, there are no widely applied business forecast or revenue models allowing for climate or weather data input.

Real-time observations and near-term forecasts are also used by tourism operators for marketing purposes. Tourism operators provide current observations (including Webcams) and short-term forecasts to promote their location or event to tourists. Other innovative marketing strategies that utilize forecast information to tailor marketing messages or target certain markets have emerged recently. For example, a major golf course operator in South Carolina monitors forecasts in nearby regional markets and when poor conditions are forecast, target-markets those locations via prearranged Internet marketing arrangements with two-day specials for weekend golf and pleasant weather [113].

The extent to which climate information is used by tourism operators for strategic, operational or marketing purposes has not been evaluated sufficiently to determine its economic value or the degree to which user needs are being met. One exception was a study of the use and need for coastal climatology information in North Carolina [73]. This study revealed that almost all the interviewees responded that marine and weather information are important in their decision-making processes, but that few managers used climatological information for longer timescales (month or longer) and that tailored products would serve as an important education resource for teaching coastal managers the utility of the long-term perspective in climatology.

Government agencies

Government agencies, while primarily climate information suppliers, are also tourism end-users of climate information. Government agencies are often directly responsible for marketing their nation and specific destinations or do so in partnership with the tourism industry. Governments are also operators of tourist attractions (national parks, for example) and other key tourism infrastructure (airports and marinas, for example). Some governments are also assisting their tourism sectors to assess the risk posed by the impacts of global climate change (Australia's Tourism Action Plan on Climate Change, for example – see Government of Australia, 2008 [114]). In these capacities, governments utilize climate information in the same ways that were previously outlined for private sector operators and destinations.

Table 5. Applications of weather derivative contracts in the tourism sector

Tourism Operator-Destination	*Derivative Contract Details*
Island of Barbados	Weather-related Challenge: low temperatures or heavy rain Weather Derivative Protection: The Barbados Perfect Weather Guarantee, gives travellers money back when daytime temperatures do not reach 26°C or when there is more than 5 mm of rain, is a recent weather derivative backed marketing campaign to promote perfect holiday weather on the island.
City of Victoria (Canada)	Weather-related Challenge: reputation for seasonal rainy conditions Weather Derivative Protection: Offers a 'Sunshine Guarantee' that will refund travellers a set amount if they experience more than 1.25 cm of rain in one day during their holiday.
Bombardier Motor Corp (Montreal)	Weather-related Challenge: limited snowfall Weather Derivative Protection: Promised buyers of new snowmobiles that if snowfall in their area was less than 50 per cent of a three-year average, they would receive a set payment.
Corney & Barrow (Wine Bar Chain, London) (Similar contracts were arranged by Massive Ltd which operates 26 Pubs and Restaurants in London)	Weather-related Challenge: low temperatures Weather Derivative Protection: Contract paid the company for every Thursday and Friday that temperatures did not reach over 24°C. The lower the temperature, the greater the payout.
Flagstaff Nordic Center Guaranteed Season Passes (Arizona)	Weather-related Challenge: lack of snow or cold temperatures required for snowmaking Weather Derivative Protection: Purchase of seasonal weather protection for season pass holders and weekend protection for special events against inadequate precipitation (less than two inches of liquid precipitation, approximately two feet of snow) between 23 November and 23March.
PGA Championship: Greater Hickory Classic at Rock Barn, North Carolina	Weather-related Challenge: rain interruption of special event Weather Derivative Protection: Purchase of derivative protection against rain event. Payment for rain accumulation over ¾ inches during the tournament.
Taste of Antwerp (Belgium)	Weather-related Challenge: rain interruption of festival with no advance booking of tickets available Weather Derivative Protection: Organizers will receive a payment should rainfall during the event exceed 9.7 mm.
Priceline.com "Sunshine Guaranteed" Vacation	Weather-related Challenge: rain during holiday tour Weather Derivative Protection: Priceline.com offered consumers a "Sunshine Guarantee" that refunded travellers if it rained Europe (>0.50 inches per day) during half or more of their vacation days at over 100 destinations in the U.S., the Caribbean, Canada and Europe.
Itravel2000.com "Let it Snow!" Weather Promotion	Weather-related Challenge: snow related travel delay Weather Derivative Protection: itravel.2000 offered consumers a refund for travel if it snowed 5 inches or more at Calgary, Halifax, Montreal or Toronto airports on New Year's Day, 1 January 2008.

The regulatory function of government and its central role in emergency preparedness and disaster planning also have important requirements for climate information. A number of regulatory frameworks that are relevant to tourism utilized historic climate information as well as near-term and occasionally seasonal forecasts: coastal management plans and setback requirements; building design standards (for heating and cooling or hurricane force winds, for example; emergency management (tourist warning systems, evacuation plans, for example); environmental impact assessments (influencing property developments as well as operations like snowmaking and desalination plants, for example); wildlife management (fish and game quotas, for example); water quality standards (swimming bans, for example); and wildfire management (open fire bans in parks and campgrounds, for example).

Knowledge gaps and needs assessment

Although climate information is increasingly available to tourism users in most countries in a number of forms, there has been little systematic assessment of the extent to which climate information is used within the tourism sector at regional, national or global scales or with any specific tourism subsector or tourist market segment. Altalo and Hale [71] and Gamble and Leonard [73] are the only known studies that have explored the climate information needs of specific tourism users; these studies provide the basis for further work in the sector.Although decision-makers in the tourism sector are increasingly aware of climate risks, no doubt in large measure due to increased concerns over global climate change, tourism operators and authorities appear generally ill-equipped to make effective use of available climate information. This review has identified major knowledge gaps with respect to: the level of end-user awareness of different types of climate information and services, exactly what climate information (basic or specialized) is being utilized by tourism end-users, how this information is being integrated into the specific decision-making processes of tourists or the tourism industry and whether there are climate information needs for end-user decision-making that are not being met with existing climate information. There also exists a tremendous need to evaluate the value of climate information for decision-making in the tourism sector both for tourists and business-destination planning, marketing and operations. This review found the following limitations in the provision of weather and climate related services to hinder the effective use of climate information in the tourism sector.

Climate information availability and resolution

Effective decision-making requires the right climate information not only in terms of the quality of the data, but in terms of its applicability to tourism operators and specific activities pursued while travelling. As discussed in

Section 2.1, historical climate information has often been unavailable for tourism planning because meteorological networks are not present (or have lacked electronic archiving) or do not adequately represent the climatic conditions in specific tourism destinations. Climate information in regions with spatially coarse meteorological networks can have reduced applicability to decision-making in the tourism sector, because activity patterns and tourism operations are highly localized, often in microclimate conditions (for example, mountain valleys, high elevations, coasts or small islands). In this review of the types of climate information provided by destinations or tourism operators, it was observed in a number of cases that the climate station operators or destination promoters used were more than 100 km away from the destination. Weather information being communicated to potential users may therefore differ substantially from the prevailing local conditions.

Recommendation: Investment is necessary to strengthen climate monitoring networks in areas where the tourism sector is vital to local economies, specifically rural areas and many developing countries (particularly Small Island Developing States), in order to improve climate risk management and climate change adaptation in the tourism sector.

Recommendation: With the risk in developing countries of permanent loss of historical climate data, which has potentially high value for managing climate risk and informing climate change adaptation, action is urgently needed establish a coordinated international data rescue initiative.

Issues of availability and representativeness of climate data for tourism destinations are likely to encumber climate change adaptation and the further development of weather-linked securities within the tourism sector. It is almost impossible to develop an effective climate change adaptation strategy without knowledge of past climate and how this baseline is projected to change in the future. The pricing of weather derivative contracts and the right of a customer to receive payment are based on weather statistics and observations at individual weather stations with a high quality historical record, and are not contingent upon conditions at a specific business location (for example, ski area, golf course) or event (for example, music concert). The more coarse the meteorological network, the less representative the conditions at the nearest weather station may be, depending on distance and local geography. The lack of sufficient weather stations and availability of adequate data reduces the potential application for weather-risk reduction through weather derivatives and weather index insurance. These issues are particularly acute in rural areas and some developing nations where tourism is a major part of the economy and that could benefit significantly from weather risk management products.

In situations where multiple sources of climate information were available for a destination (NMSs, private meteorological company, tourism operators), inconsistent information was sometime being provided to tourists. Current conditions more than 10°C to 15°C apart were occasionally reported, as were very different forecasts for temperature, rain and winds. Such varied reporting of climate conditions and forecasts is likely to raise questions about the validity of the source among tourists and will impact tourist decision-making for short-term visits (day trips and weekend travel especially) and highlights the importance of forecast accuracy for some segments of the tourism sector.

Recommendation: Strengthening of climate monitoring networks is necessary to support the development and access to innovative financial products (weather derivatives and index insurance) to manage climate risk in the tourism sector.

Climate information represents a double-edged sword for the tourism sector. While accurate climate information can be invaluable to the tourism industry, inaccurate climate information that deters visitation is a lament heard often by all of the experts consulted in this review and has been the subject of several workshop discussions and media stories. For example, at the Climate, Weather, and Tourism Workshop in North Carolina [115], multiple tourism stakeholders voiced frustration with media reporting of weather and how it unnecessarily affected tourism. Wineries in the region saw visitation decline for an entire season after inaccurate reports that spring frost had wiped out that year's crop [116]. Similar concerns about forecast skill and media coverage of weather forecasts were expressed by other tourism operators and tourism authorities in the region, especially the typical margin of error in early hurricane track forecasts (7–10 forecasts) that have been observed to be very damaging to the tourism economies of regions at the low probability edges of the forecast track [117]. The accuracy of early forecasts of extreme events was also an important topic of discussion at the Secure and Sustainable Living – Social and Economic Benefits of Weather, Climate and Water Services Conference in 2007 (see statement below). Seasonal forecasts have also recently been shown to have adversely affected travel decisions in the United Kingdom. In 2009, travel agents and tour operators observed that following the NMS long-range summer forecast of "unusually warm, dry weather with heatwaves up to 30°C", demand for foreign holidays declined substantially [44]. After a very rainy month of July and a revised forecast for "wet weather until September", the Association of British Travel Agents reported an increase in travel bookings of up to 40 per cent and a diminished supply of package holidays to sunshine destinations [44].

"...the very early warning of extreme weather events, such as tropical cyclones, can unwittingly put off tourists from destinations unlikely to be affected by the event or if impacted, well outside the time that the tourist would be in the destination. Thus there is a need for the industry and forecasters to work closely together to resolve the issues that arise from lack of specificity in forecasts, while ensuring public safety."

Final Communiqué Secure and Sustainable Living –

Social and Economic Benefits of Weather, Climate and Water Services

Madrid, 19-22 March 2007

The article "Gold Coast Businesses Unhappy with Weather Forecasting" (Nolan, 2001) describes the concerns of several tourism operators in Queensland, Australia, about the accuracy rate of the local forecasts and the financial impact on their businesses. Some tourism operators have indicated that one of the reasons they included additional sources of localized climate information and live Webcams on their Website, was to overcome perceived limitations in forecasts and biased reporting by mass media, who, in their words, were often "looking for a story" about a storm or poor travel conditions. As the statement below indicates clearly, tourism businesses affected by weather and forecasts have an economic stake in the quality of the climate information.

"It's a fair assumption to suggest that the weather and the weather forecast – however derived – is a very important part of the (tourist) decision making process. ... I think we both (NMS and tourism industry) want accuracy, but also at the same time, we want it balanced, with a situation where the travelers are not deterred because the off chance of a shower is portrayed as if it was a certainty."

Daniel G-Schwind Chief Executive, Queensland Tourism Industry Corporation

(in Nolan, 2001)

Unsubstantiated speculation about the impacts of climate change on tourism destinations is likewise problematic and may adversely affect investment and eventually visitation needlessly. A number of media stories have foretold the major threat that increased future summer temperatures pose for tourism in the Mediterranean. Indeed some have gone so far as to state that, "The likelihood [is] that Mediterranean summers may be too hot for tourists after 2020" (Guardian, 2006 - based on Amelung and Viner [118]) and that "by 2030, the traditional British package holiday to a Mediterranean beach resort may be consigned to the 'scrapheap of history'" [119]. Such pronouncements have been shown to be unfounded scientifically (see Scott et al., [74] and Rutty and Scott, [101]). Similar speculation about the demise of winter sports tourism in the European Alps, North America, and other ski regions is also widespread, with some stories going as far as identifying specific ski areas expected to be lost and "where not to buy ski properties" (real estate investment).

Unfortunately, such misinformation continues to be perpetuated by other travel writers even years after the initial stories have been published (see Osborne, [120]; Munns, [121]; *The Independent*, [122]; Couttswoman, [123]; Newsom, 2007) and will continue to negatively impact the reputation of certain destinations. This type of speculation diminishes the credibility of all other climate change assessments designed to help the tourism sector and destination communities adapt to future climate change. In the same way accurate forecasts are in the best interest of the tourism sector, so too are accurate representations of the risks posed by climate change. The tourism sector, particularly destinations, need to work closely with informed scientists to accurately understand the risks and opportunities posed by climate change and ensure that the misrepresentations of climate change vulnerability in the media do not go unanswered.

Recommendation: The development of regionally and locally specific climate change scenarios is necessary to facilitate effective climate change adaptation by the tourism industry and tourism-dependent communities. The refinement of near-term climate change predictions (covering the next 25–30 years) that are most relevant to business investment and government policy timeframes are particularly encouraged.

Influence of climate information on decision-making

Knowledge about the process of how tourism end-users integrate weather and climate information into specific decisions remains very limited. Key knowledge gaps remain regarding the decisiveness of weather and climate as a factor in decision-making and when end-users require climate information in different decision-making processes. The proportion of tourists travelling primarily for climate-related motivations or to engage in climate-sensitive activities remains unknown and central to understanding the relative climate sensitivity of destinations. Describing the decision-making process requires, at a minimum, describing the key decision- makers, their goals and the context in which they operate (the decision-making environment), the information they use to make decisions, the alternative actions available to them, and the important decision points [124][125]. Climate information is typically embedded in a matrix with other relevant information [27][59][70][124] and disentangling the role of climate and its relationship to other situational factors in major decision-making processes in tourism remains an important objective for future research.

Tourists experience and respond to the integrated effects of the weather elements – thermal, physical, aesthetic [70] that comprise climate [70][126][127][128]; however, there remains a very incomplete understanding of the relative importance of different climate parameters to tourists, the range of climate preferences among tourists in terms of the optimal conditions and thresholds of unacceptability for certain variables, how preferences differ for specific tourism environments or destinations and whether significant cultural, regional or market segment differences exist in climate preferences [43][95][128]. The complexities of tourist climate preferences are only beginning to be examined. For example, recent survey research in temperate regions of North America, Europe and New Zealand [38][43][101][128] has found that the importance of key climate variables as well as perceptions of optimal conditions and thresholds for unacceptable conditions differed substantially among tourists and in specific tourism environments (Table 6 and Figures 21 and 22). More research is required to understand whether similar climate preferences exist among tourists from other climatic regions of the world as well as to understand the specific climate sensitivity of specific tourism activities.

Table 6. Importance of weather variables in different tourism environments

Importance Rank	*Beach*	*Mean (1–7)*	*Urban*	*Mean (1–7)*	*Mountain*	*Mean (1–7)*
1	Sun	6.14	Temp	5.98	Rain	6.04
2	Temp	6.11	Rain	5.77	Temp	5.84
3	Rain	5.87	Sun	5.14	Sun	5.55
4	Wind	5.13	Wind	4.75	Wind	5.41

Source: Scott et al. [128]

Temp = "comfortable temperature"; Rain = "absence of rain"; Wind = "absence of strong wind"; Sunshine = "presence of sunshine"

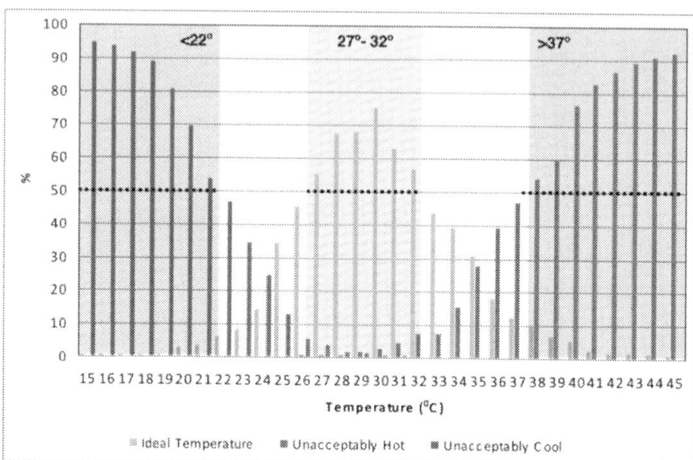

Figure 21. Tourist rating of temperatures for beach holidays (Source: Rutty and Scott [101])

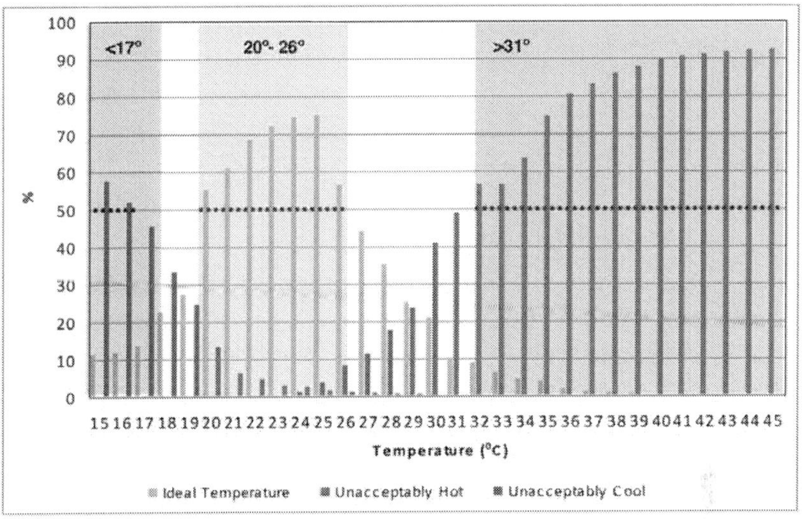

Figure 22. Tourist rating of temperatures for urban holidays (Source: Rutty and Scott [101])

Recommendation: Support is necessary for the fundamental multidisciplinary research needed to understand the salience of climate (both in source markets [push factor] and destinations [pull factor]) in different travel decision-making contexts, cross-cultural climate preferences for major destination types, the effect of weather on holiday satisfaction and future travel choices and the climate sensitivity of major tourism activities.

Fundamental knowledge of climate conditions that strongly influence destination choice and activity satisfaction is a prerequisite to the development of specialized climate products for the tourism sector. Without it, the validity of existing indices, and those utilized in the research literature (the Tourism Climate Index by Mieczkowski [126], for example) remain uncertain. Unfortunately, most of the newly developed specialized products from private meteorological companies lack transparency in the methods and data sources to properly evaluate their application to either domestic or international travellers. Many specialized products will serve to inform destination and activity choices, and their design must be transparent to the consumer to allow objective destination comparisons and marketing claims in a global tourism marketplace. The "Heat Wave Vacation" story in the *Wall Street Journal* [129] illustrates how some travellers feel misled by travel operators and destination marketers about the climate of destinations and how they would benefit from a tool to compare climatic conditions of destinations they are considering. Some specialized products will also affect decisions that have implications for the personal safety of travellers and therefore the design and data inputs should be transparent to tourism professionals that might use them as well as to the travellers themselves.

Recommendation: Developers of specialized climate products for the tourism sector, whether the private sector, universities or governments, are encouraged to disclose the scientific methodology or market testing results to demonstrate validation in the tourism marketplace.

Recommendation: The tourism sector, in collaboration with NMSs, private meteorological companies and university researchers, are encouraged to develop accepted standards for specialized climate products, to ensure consistent and accurate communication of climate information to international travellers and to facilitate objective destination comparisons and marketing claims in a global tourism marketplace.

The availability of climate information has not been the sole barrier to understanding the interface of climate and tourism sufficiently to develop specialized climate services, facilitate greater involvement in the emerging weather derivatives and index insurance market or determine the weather and climate information for tourism decision-making. Access to tourism data has been a frequently noted impediment to research and the development of specialized climate services and climate change assessments [41][42][45][130][131][132][133] and a common frustration of many of the experts consulted for this review. For this partnership to move forward and succeed, the tourism community must be an active contributor. Furthermore, the importance of collecting accurate and consistent domestic and international tourism data to advance scientific understanding of the climate–tourism interface and the implications of climate change cannot be overemphasized. Past recommendations of the UNWTO [5][134] and the broader tourism research community [135] for continued efforts to maintain and enhance monitoring of tourism performance indicators are echoed here.

Recommendation: Collaboration between governments, universities, communities and the private sector (tourism businesses, meteorological service companies, financial services), must be strengthened to drive innovation that connects climate information to the needs of the tourism sector and tourism-dependent communities.

Recommendation: The active collaboration of the tourism industry is necessary to support the development of climate services to improve outcomes for the sector, and the industry is strongly encouraged to provide increased access to sectoral data, consult on specific climate information needs and constraints to its use, provide expert review of specialized products and create effective strategies to communicate weather and climate information to tourists.

Socio-economic benefits of weather and climate information for tourism

Effective use of climate information has the potential for avoiding injury and death, averting property and environmental damage and a wide range of other societal benefits [124]. There is a growing literature on the economic value of climate information and forecasts [136] as well as techniques to valuate non-market benefits. The tourism sector is virtually absent from this literature, although Altalo and Hale [71] contend that the financial benefits of weather and climate information for the sector are likely to be very substantial. One of the few attempts to evaluate the specific financial benefits of improved weather and climate information for the tourism sector was a NOAA study of the multisector benefits of new observational equipment [137][138]. Specific to the tourism sector, the study found that weather information and hurricane forecasts from the new satellite imager and sounder would create US$ 196 million per year in socio-economic benefits through improved golf safety, irrigation efficiency, grounds maintenance, tournament and personal golf planning [138] as well as US$ 31 million per year in economic benefits, from damage avoidance in recreational boating, amusement and recreation services [137][139].

Moreover, there are no studies of the financial or non-market benefits of specialized products for tourism or the willingness to pay for climate information among tourists and tourism operators. Consequently, the value of climate information to tourists and the tourism industry remains unknown. Hypothetically, if tourists were willing to pay 1 Euro for weather and climate information needed for trip planning for each of the estimated 900 million international and 8 billion domestic trips, then the global value of climate services would be very significant even without consideration of its value for supply-side operations and marketing. Based on the success of studies that have applied techniques for economic and social valuation of climate services (market prices, normative market models, descriptive behavioural response studies and contingent valuation studies, for example) in other economic sectors, it is clear that uncovering the potential value for the tourism sector and determining how to fully realize that potential remain critical areas for future inquiry.

Recommendation: An interdisciplinary initiative should be established to evaluate the economic and non-market societal value of climate information for decision-making by tourists and tourism operators.

Communicating weather and climate information to the tourism sector

The production of climate information alone is not sufficient for travel or business planning and decision-making. Information must be delivered to end-users in a form that is relevant to them and that they have the capability to interpret. As identified in Section 2.1, there are a range of communication channels for the delivery of climate information to tourists and the tourism sector. How climate information is communicated to tourists and tourism subsectors is largely unexplored and communication channels, especially related to warnings of abrupt and dangerous weather events, are not widely documented within the tourism sector. There has been almost no evaluation of what sources of climate information tourist or tourism operators utilize, nor the effectiveness of different communication pathways and formats.

Gamble and Leonard's [73] study is the only known attempt to identify Web-based tourist preferences for climate information. Examining the communication of coastal climatology information, the authors found that simple sites were perceived by tourists to be most effective and included: (a) a limited amount of information presented on the Website; (b) efficient and clear Web navigation features; (c) limited use of scientific jargon and graphics; and (d) limited use of colours and flashy graphics. Also of note was the identified need for more local data and the annoyance with the need to purchase or download software to further assess climate data. Overall, the authors found that a more organic, bottom-up Website design that reflects the needs of users and locals was beneficial. There is a clear need to test these findings for universal applicability across the breadth of tourism-related climate information products available to end-users.

Most weather forecasts today still contain little uncertainty information. Without this information, forecasts can easily be misinterpreted and potentially misused in decision-making. This can have potentially negative social and economic consequences (especially with respect to hazardous weather events). Although this challenge is recognized by many NMSs and private meteorological companies, overall the meteorological community has a limited understanding of how to communicate weather forecast uncertainty effectively to users with diverse technical and intellectual capabilities [140]. Interdisciplinary research on tourist and tourism sector perceptions of uncertainty within forecasts, preferences for uncertainty communication formats and a better understanding of the decision stakes related to uncertainty across a wide set of weather and climate information end-users is needed.

Recommendation: Greater effort should be made to consult with major tourism end-users about their needs for climate information. This consultation must be done regionally in order to adequately represent specific information needs and the capabilities of regional providers.

The responsibility for ensuring that the natural hazard information is communicated to the public rests squarely with government agencies. Ensuring adequate community response requires a close partnership and well established and regularly tested working arrangements and channels of communication. However, understanding how weather forecasters and emergency managers use climate information and how this information is communicated to the tourism sector is currently limited.

Social science research conducted during the last decade has highlighted the difficulty in preparing visitors to effectively respond to warning messages [141][142][143][144]. Tourists can be particularly vulnerable to natural hazards because they often visit highly dynamic environments, can have limited familiarity with the places they are visiting and with common meteorological hazards and remote locations can lack communication channels for public warnings of impending hazards. The tornado that struck the Pine Lake, Alberta, (Canada) campground in July 1999, killing nine and injuring over 100, is one such example. Although the Meteorological Service of Canada had issued a tornado warning well in advance of the event, there was a communication breakdown, and virtually all of the tourists at the campground were unaware of the warning of imminent severe weather and there was no on-site emergency warning system available (a siren system used in urban areas, for example). Furthermore, because tourists can be unfamiliar with the local language and are less likely to utilize local media information sources (either purposively or unintentionally), tourists are less apt to receive hazard warnings when they are issued, and prompt communication with them of imminent climatic hazards poses a particular challenge [85].

Recommendation: An interdisciplinary evaluation of best practices for communication of climate information, particularly specialized products and forecast uncertainty, to tourism end-users is encouraged.

CONCLUSIONS

The past decades have seen tremendous advances in climate science and knowledge of how the global climate is likely to change over the twenty-first century as a result of anthropogenic greenhouse gas emissions and feedbacks in the global climate system. The revolution in communication technology, particularly the Internet and more recently mobile personal data devices (Smartphones, for example) has also revolutionized the climate information available to tourists and the tourism industry. Although interest in the climate–tourism interface has also increased markedly during the past decade, as evidenced by a doubling in the number of scientific publications on tourism and climate between 1996–2000 and 2001–2005 [145], the research community has yet to evaluate how the revolution in climate information and information communication technologies has translated into improved decision-making in the tourism sector.

This review has documented the many sources of climate information now available to the tourism sector and the continuing development of specialized climate products for tourists and tourism operators. It also revealed that there has been no systematic evaluation of the extent and nature of climate information use in any tourism subsector or specific destination region. There is very little insight into the role of climate information in specific decision processes within the tourism sector (either demand or supply- side), the economic and non-market value of climate information and related service to end-users and society or the most effective ways to communicate climate information to diverse tourism end-users. Addressing these key knowledge gaps and others identified throughout this review would provide significant opportunities to enhance decision-making and reduce climate risk in the sector.

Critically, accumulating evidence indicates that climate change, particularly high emission scenarios, will be a pivotal issue affecting the medium- and long-term future of tourism development and management [27][66][67][68]. Consequently, it is recognized that the need for climate services will increase throughout the twenty-first century as the magnitude of climate change increases and the ability to rely on previous experience diminishes. As the statements from the Secretaries-General of the WMO and UNWTO indicate, both organizations recognize that improving the use of climate information is an important strategy to facilitate sustainable development of tourism that contributes to the United Nations Millennium Development Goals.

> "Given that climate change is expected to pose an increasing threat to tourism operations in many destinations . . ., WMO urges governments and the private sector to increasingly use climate information . . ., and to take additional steps towards incorporating climate considerations in tourism policies, development and management plans."

WMO Secretary-General M. Michel Jarraud (2007)

> "Climate change will constitute an increasing risk for tourism operators in many destinations. With many tourism activities heavily dependent on the climate and insurance policies increasingly affected by natural hazards,

accurate weather information and forecasting of extreme climatic events are becoming ever more important for tourism businesses."

Improving the use of climate information in the tourism sector is a challenge that will require closer collaboration among the climate and tourism research communities (both physical and social scientists), NMSs, government tourism authorities and the tourism industry. Over a decade ago Smith [60] argued that the limited research on the complex interactions of climate and tourism was in large measure because, "... meteorologists and leisure specialists rarely communicate with each other." With substantial opportunities to more effectively deliver climate services that are of immediate and recognizable value to tourists and the tourism industry and the imperative of adapting to climate change in the decades ahead, the time has come for the conversation between climate service providers and tourism end-users to begin in earnest. Positively, a number of key new partnerships have emerged in recent years that provide a solid foundation for necessary future collaboration. The collaboration between the WMO and UNWTO to establish an Expert Team on Climate and Tourism in 2005 is a critical initiative at an international level. New partnerships have also developed between meteorological institutions and tourism stakeholders [146]. This cooperation has taken multiple forms, from new forecasts for tourism destinations, improved media training and cooperation to deliver forecasts related to tourism, to specific contracts between meteorological services and destinations, tour operators and other stakeholders.

What is now most needed is a strategic initiative that will encourage direct interaction and catalyse regional and local partnerships among climate service providers, governments, universities and the diverse user groups within the tourism sector. It is recommended that a series of multi-objective, capacity-building workshops be undertaken over the next five years in major tourism regions with diverse climate sensitivities. The regional workshops would build capacity in both the climate services and tourism sectors. The workshops would stimulate interest in the impacts of climate variability and change, increase awareness of existing basic and specialized climate services among tourism professionals, determine specific needs for climate forecasts and information by tourism subsectors and provide educational sessions in the use of probability-based products, decision support tools and weather risk markets and weather derivative/index insurance products.

Despite visibly increased attention to the challenge of climate change by the UNWTO over the last five years, a number of studies of climate change awareness and adaptation practice for the UNWTO, the United Nations

Environment Programme (UNEP) and WMO [74] and a recent multisectoral comparison by KPMG [147] consistently found low awareness of climate change risk and little evidence of strategic planning within the tourism industry. Developing awareness of the regional, tourism-specific implications of climate change, with the aim of improving adaptation, would therefore be a second major objective of the capacity-building workshops. Climate services professionals would be provided with feedback on the diverse information needs of specific tourism stakeholders in their region, the potential development of specialized products and how to improve communications to enhance the utility of climate information.

The lessons learned from Regional Climate Outlook Forum workshops undertaken by WMO, the Awareness and Preparedness for Emergencies at the Local Level (APELL) workshops for tourism destinations coordinated by UNEP, and other recent climate and tourism specific workshops (Caribsave in the Bahamas and Jamaica, GEF and UNWTO in Fiji, NOAA and East Carolina University in the United States), would inform the themes and structure of the workshops. In addition to establishing key regional partnerships, these workshops would build a pool of professionals qualified to transfer advances in climate science and climate prediction to climate-related decision frameworks and develop decision-support tools for the tourism sector. By adopting a regional approach, the capacity-building workshops would also overcome some of the major regional knowledge gaps on the sensitivity of tourism to climate variability and change in developing nations, notably Africa, South-East Asia and Latin America [74][148].

Recommendation: A series of multi-objective, capacity-building workshops should be initiated in major tourism regions around the world, in order to foster the direct interactions and partnerships between climate service providers and tourism user groups needed to make significant progress in the application of climate information in the tourism sector.

Recommendation: Training the next generation of tourism professionals to utilize climate information to reduce climate risks and adapt to climatic change in the decades ahead is a priority, and it is urged that a Climate Risk Management training module be created for use by tourism and hospitality schools around the world.

As an important contributor to national and local economies around the world, tourism is also highly interlinked with other major sectors for which white papers are being produced, such as health (tropical disease outbreaks and dispersion by travellers, for example), urban (heat warnings, for example) and biodiversity (involvement in coral reef monitoring programs, for example). Consequently, common interests and recommendations are probable, providing a basis for multisectoral collaboration on the improvement of global climate services that foster sustainable development.

ACKNOWLEDGEMENTS

The authors wish to acknowledge the help of the following experts in the preparation and review of this paper: Bruno Abegg, University of Zurich, Switzerland; Susanne Becken, Lincoln University, New Zealand; Luigi Cabrini, United Nations World Tourism Organization; Jean Paul Ceron, Université de Limoges, France; Christopher de Frietas, University of Auckland, New Zealand; Stefan Gössling, Lund University, Sweden; Western Norway Research Institute, Norway; Michael Hall, University of Canterbury, New Zealand; Brenda Jones, Parks Canada; Jacqueline Kuehnel, JK Consulting and Enterprises; Patrick Long, East Carolina University, United States; Leslie Malone, World Meteorological Organization; Geoff McBoyle, University of Waterloo, Canada; Alvaro Moreno, Maastricht University, Netherlands and Neville Trotz, Caribbean Climate Change Centre.

REFERENCES

1. United Nations World Tourism Organization (UNWTO), NWTO technical manual: Collection of Tourism Expenditure Statistics, 1995.http://pub.unwto.org/WebRoot/Store/Shops/Infoshop/Products/1034/1034-1.pdf.(Retrieved on 2009-07-11).

2. United Nations World Tourism Organization (UNWTO), World Tourism Barometer – June 2009. http://www.unwto.org/facts/eng/pdf/ barometer/UNWTO_Barom09_2_en_excerpt.pdf (Accessed 15 July 2009).

3. United Nations World Tourism Organization (UNWTO), Tourism 2020 Vision, Madrid, UNWTO, 2001.

4. World Travel and Tourism Council, Tourism impact data and forecasts, 2009. http://www.wttc.org/eng/Tourism_Research Tourism_Economic_Research (Accessed 1 July 2009).

5. United Nations World Tourism Organization (UNWTO), Workshop on Tourism Statistics, 2008. http://www.sesrtcic.org/imgs/news/Image/ REPORT_ ANKARA_ TURKEY_JUNE%202008.pdf.

6. United Nations Conferences on Trade and Development (UNCTAD), UNCTAD Handbook of Statistics 2008, New York and Geneva, United Nations, 2008.

7. United Nations World Tourism Organization (UNWTO), Compendium of tourism statistics (2007 edition), Madrid, UNWTO, 2007.

8. Wall, G. and C. Badke, Tourism and climate change: an international perspective. Journal of Sustainable Tourism. 2 (1994) 193.

9. Burton, R., Travel Geography, Essex, Longman, 1995.

10. Boniface, B. and C. Cooper, The Geography of Travel and Tourism, London. Butterworth-Heinemann. 1994.

11. S. Pike, Destination image analysis – A review of 142 papers from 1973 to 2000. Tourism Management. 23 (2002) 541.

12. P. Anderssen and R. Colberg, Multivariate analysis in travel research: A tool for travel package design and market segmentations. Proc. Fourth Annual Travel Conference of the Travel Research Association, Travel Research Association, Sydney, Australia, 1973.

13. Hu, Y.and J. Ritchie, Measuring destination attractiveness: A contextual approach. Journal of Travel Research. 32 (1993) 25.

14. Butler, R., Seasonality in tourism: Issues and implications, in: Seasonality in Tourism (T. Baum and S. Lundtorp, eds), London. Pergamon, 2001.

15. IUCN, Forest fires in the Mediterranean – backgrounder, IUCN, Gland, Switzerland, 2007. http://www.uicnmed.org/web2007/documentos/Background_ med_forest _fires.pdf (Accessed 1 July 2009).

16. Butler, A., Tourism burned: Visits to parks down drastically, even away from flames. Rocky Mountain News, 15 July, 2002.

17. Associated Press, Royal Gorge tourism hurt by fires, drought, 3 September, 2002a.

18. Associated Press, Rough year for rafters, 3 September, 2002b.

19. United States Environmental Protection Agency, Global Climate Change: What Does it Mean for South Florida and the Florida Keys, Report on the Environmental Protection Agency Public Consultations in coastal cities. May 24–28, Washington, Environmental Protection Agency, 1999.

20. Pack, T., Florida tourism problem, Hotel Online – News for the Hospitality Executive, 2004. www.hotel-online.com/News/PR2004_4th/Oct04_FloridaNextYear.htm (Accessed 28 July 2005)

21. Deravi, M. and P. Smith, Economic impact Alabama travel industry 2004, Montgomery, Alabama Bureau of Tourism and Travel, 2005.

22. USA Today, Florida's hurricanes boost Arizona tourism, 2005. www.usatoday.com/travel/news/ 2004-10-11-az-hurrican_x.htm (Accessed 29 July 2005).

23. Williams, D., Wilma deals $800 million blow to Mexico travel industry, 2005. www.cnn.com/2005/travel/10/26/wilma.mexico.travel/index.html (Accessed 27 October 2005).

24. Bhatnagar, P., Uncertainty and fear to rebuild: Mississippi officials say casino companies are hesitant to rebuild, state could lose billions. CNN Monday, 6 September, 2005. http://money.cnn.com/2005/09/05/news/economy/katrina_business/.

25. World Meteorological Organization (WMO), Technical Conference – Climate as a Resource. Beijing, China, 1–2 Nov. WMO, Geneva. 2005.

26. United Nations World Tourism Organization (UNWTO), Emerging Tourism Markets – The Coming Economic Boom. Press Release, UNWTO Madrid, 24 June 2008.

27. Scott, D., Climate change and sustainable tourism in the 21st century, in: Tourism Research: Policy, Planning, and Prospects (J. Cukier, ed.) Waterloo, Department of Geography Publication Series, University of Waterloo, 2006.

28. Dawson, J., D. Scott and G. McBoyle, Analogue analysis of climate change vulnerability in the US northeast ski tourism. Climate Research. 39 (2009) 1.

29. Goodman, J., Battered ski area sweating for snowfall. Seattle Times. 27 October, 2005.

30. National Ski Areas Association, Kottke National End of Season Survey 1996/97. National Ski Areas Association, 1997.

31. National Ski Areas Association, Kottke National End of Season Survey 1999/00. National Ski Areas Association, 2000.

32. National Ski Areas Association, Kottke National End of Season Survey 2004/05. National Ski Areas Association, 2005.

33. Jones, B., D. Scott and H. Abi Khaled, Implications of climate change for outdoor event planning: A case study of three special events in Canada's Capital region. International Journal of Event Management. 10 (2006) 63.

34. Scott, D., G. Wall and G. McBoyle, The evolution of the climate change issue in the tourism sector, in: Tourism, Recreation and Climate Change (M. Hall and J. Higham, eds), London, Channelview Press, 2005a.

35. Getz, D., Why festivals fail. Event Management. 7 (2002) 209.

36. Weather Bill, Inc., Impact of Climate Change on Golf Playable Days in the United States, 2007. http://www.weatherbill.com/assets/LandingPageDocs/golfstudy.pdf (Accessed 11 June 2009).

37. Mintel International Group, Special report - holidays. Leisure Intelligence. London, Mintel International Group, 1991.

38. Lohmann M. and E. Kaim, Weather and holiday preference - image, attitude and experience. Revue de Tourisme. 2 (1999) 54.

39. Ontario Ministry of Tourism and Recreation, If the future were now: Impacts of aging in the Canadian market on tourism in Ontario. Toronto, Ontario Ministry of Tourism and Recreation, 2002.

40. Kozak, M. Comparative analysis of tourist motivations by nationality and destinations. Tourism Management. 23 (2002) 207.

41. Hamilton, J.M. and M.A. Lau, The role of climate information in tourist destination choice decision-making, in: Tourism and Global Environmental Change (S. Gössling and C.M. Hall, eds) London, Routledge, 2005.

42. Gössling, S., M. Bredberg, A. Randow, P. Svensson and E. Swedlin, Tourist perceptions of climate change: A study of international tourists in Zanzibar. Current Issues in Tourism. 9 (2006) 419.

43. Moreno, A., Mediterranean tourism and climate (change): A survey-based study, in: Proceedings of 7th International Symposium on Tourism and Sustainability, Travel and Tourism in the Age of Climate Change. University of Brighton, Brighton, England, July 8-10, 2009.

44. Hill, A., Holiday deal abroad vanish in rush to flee the rain. The Observer. 9 August, 2009.

45. Smith, K., Tourism and climate change. Land Use Policy, April, 1990, 176.

46. AgnewTourism, M., in: Economic Impacts of the Hot Summer and Unusually Warm Year of 1995 (J. Palutikof, S. Subak and M. Agnew, eds), Department of the Environment Report, Norwich, 1995

47. Giles, A. and A. Perry, The use of a temporal analogue to investigate the possible impact of projected global warming on the UK tourist industry. Tourism Management. 19 (1998) 75.

48. Agnew, M. and J. Palutikof, Impacts of short-term climate variability in the UK on demand for domestic and international tourism. Climate Research. 31 (2006) 109.

49. Bigano, A., A. Goria, J. Hamilton and R. Tol, The Effect of Climate Change on Extreme Weather Events and Tourism. Working Paper of Fondazione Eni Enrico Mattei, Milano. 2005.

50. Jorgensen, F., and G. Solvoll, Demand models for inclusive tour charter: The Norwegian case. Tourism Management. 17 (1996) 17.

51. Wilton, D. and T. Wirjanto, An analysis of the seasonal variation in the national tourism indicators. Ottawa., Canadian Tourism Commission. 1998.

52. Paul, A., Weather and the daily use of outdoor recreation areas in Canada, in: Weather Forecasting for Agriculture and Industry (J. Taylor, ed.), Newton Abbot. David and Charles Publishers. 1972.

53. Meyer, D. and K. Dewar, A new tool for investigating the effect of weather on visitor numbers. Tourism Analysis. 4 (1999) 145.

54. Hamilton, L., D. Rohall, B. Brown et al., Warming winters and New Hampshire's lost ski areas: An integrated case study. International Journal of Sociology and Social Policy. 23 (2003) 52.

55. Jones, B. and D. Scott, Climate change, seasonality and visitation to Canada's National Parks. Journal of Parks and Recreation Administration. 24 (2006) 42.

56. Scott, D. and B. Jones, A regional comparison of the implications of climate change of the golf industry in Canada, The Canadian Geographer. 51 (2007) 219.

57. Moreno, A., B. Amelung and L. Santamarta, Linking beach recreation to weather conditions: A case study in Zandvoort, Netherlands. Tourism in Marine Environments. 5 (2008) 111.

58. Shih, C., S. Nicholls and D. Holecek, Impact of weather on downhill ski lift ticket sales. Journal of Travel Research. 47 (2009) 359.

59. Adams, R. Uncertainty in nature, cognitive dissonance, and the perceptual distortion of environmental information: Weather forecasts and New England beach trip decisions. Economic Geography. 49 (1973) 297.

60. Smith, K. The influence of weather and climate on recreation and tourism. Weather. 48 (1993) 398.

61. Charmichael, B., Conjoint analysis of downhill skiers used to improve data collection for market segmentation. Journal of Travel and Tourism Marketing. 5 (1996) 187.

62. König, U. and B. Abegg, Impacts of climate change on tourism in the Swiss Alps. Journal of Sustainable Tourism. 5 (1997) 46.

63. Bicknell, S. and P. McManus, The canary in the coalmine: Australian ski resorts and their response to climate change. Geographical Research. 44 (2006) 386.

64. Williams, P., K. Dousa and J. Hunt, The influence of weather context on winter resort valuations by visitors. Journal of Travel Research. 36 (1997) 29.

65. United Nations World Tourism Organization (UNWTO), Climate Change and Tourism: Proceedings of the First International Conference on Climate Change and Tourism, Djerba, Tunisia, 9–11 April 2003. Madrid, UNWTO, 2003.

66. Gössling, S. and C.M. Hall, Uncertainties in predicting tourist travel flows based on models. Editorial Essay. Climatic Change. 79 (2006) 163.

67. Becken, S. and J. Hay, Tourism and Climate Change – risks and opportunities. Clevedon, Channel View Publications, 2007.

68. UNWTO, UNEP and WMO, Climate Change and Tourism: Responding to Global Challenges. Madrid, United Nations World Tourism Organization; Paris, United Nations Environment Program; Geneva. World Meteorological Organization, 2008.

69. G. McBoyle, Approaching tourism through climate, in: Approaching Tourism (G. Wall, eds), Department of Geography Occasional publication 21, University of Waterloo, Waterloo, 2007.

70. De Freitas, C.R., Tourism climatology: Evaluating environmental information for decision-making and business planning in the recreation and tourism sector. International Journal of Biometeorology. 48 (2003) 45.

71. Altalo, M. and M. Hale, Requirements of the US Recreation and Tourism Industry for Climate, Weather and Ocean Information. Consultants report to NOAA, 2002.

72. World Meteorological Organization (WMO), WMO questionnaire on the role and operation of National Meteorological Services: Some findings. WMO Bulletin. 51 (2002) 342.

73. Gamble, D.W. and L.A. Leonard, Coastal Climatology Products for Recreation and Tourism End Users in Southeastern North Carolina, NOAA Coastal Services Center, 2005.

74. Scott, D., B. Amelung, S. Becken, J.-P. Ceron, G. Dubois, S. Gossling, P. Peeters and M. Simpson, Technical Report, in: Climate Change and Tourism: Responding to Global Challenges. Madrid, United Nations World Tourism Organization; Paris, United Nations Environment Program; Geneva, World Meteorological Organization, 2008.

75. World Meteorological Organization (WMO), Communications Plan on PWS activities Responsibility of NMHSs in Issuing Meteorological Information, 2009. http://www.wmo.ch/pages/prog/amp/pwsp/legalstatus_en.htm (Accessed 22 June 2009).

76. Great Barrier Reef Marine Park Authority, Coral Bleaching on the Great Barrier Reef, 2007. http://www.gbrmpa.gov.au/corp_site/info_services/science/climate_change/climate_change_and_ the_great_barrier_reef/coral_bleaching_on_the_great_barrier_ reef.

77. Great Barrier Reef Marine Park Authority, Coral Bleaching Response Plan 2008-2009, Commonwealth of Australia, 2008.

78. The Weather Channel®, Major League Baseball Weather Forecast, 2009a. http://www.weather.com/activities/events/mlb/?from= secondarynav (Accessed 9 July 2009).

79. The Weather Channel®, Ski Conditions, 2009b. http://www.weather.com/activities/recreation/ ski/?from=secondary nav (Accessed 9 July 2009).

80. The Weather Channel®, Golf Course Weather, 2009c. http://www.weather.com/activities/recreation /golf/?from= secondarynav (Accessed 9 July 2009).

81. The Weather Channel®, Fishing Forecast, 2009d. http://www.weather.com/activities/recreation/ outdoors/fishing/?from=secondarynav (Accessed 9 July 2009).

82. SnowForecast.com. Resort Weather, Webcams and Ski Snow Reports. http://www.snow-forecast. com/ (Accessed 9 July 2009).

83. Surf-Forecast.com, Surf, Swell, Wind and Wave Forecasts, 2009. http://www.surf-forecast.com/ (Accessed 9 July 2009),

84. Canadian Avalanche Association (CAA), InfoEx®, 2009. http://www.avalanche.ca/CAA_CAIS_ InfoEx (Accessed 9 July 2009),

85. United Nations World Tourism Organization (UNWTO), Handbook on Natural Disaster Reduction in Tourist Areas. Published by the World Tourism Organization and the World Meteorological Organization, 1998.

86. Canadian Avalanche Association (CAA), International Danger Scale, 2009. http://www.avalanche.ca/CAC_ Bulletin_DangerScale (Accessed 9 July 2009).

87. Swiss Federal Institute for Forest, Snow and Landscape Research, 2009. http://www.slf.ch/lawineninfo/zusatzinfos/lawinenskala-europa/index_EN (Accessed 9 August 2009).

88. The Weather Channel®, The Weather Channel Interactive Introduces Premium Mobile Weather Application on Apple App Store: Ad-free App Includes TruPointSM Future Radar Maps in Motion, 2009e. http://press.weather.com/press_detail.asp?id=235 (Accessed 9 July 2009).

89. Perry, A., Recreation and tourism, in: Applied Climatology (R.D. Thompson and A.H. Perry, eds), London, Routledge,1997.

90. McBoyle, G., pers. comm. (Vice President Academic, University of Waterloo, Waterloo, Canada), 2009.

91. Jones, B., personal communication (Senior Social Scientist, Parks Canada, Ottawa, Canada), 2009.

92. de Freitas, C.R., The climate-tourism relationship and its relevance to climate change impact assessment, in: Tourism, Recreation and Climate Change: International Perspectives (C.M. Hall and J. Higham, eds), Channelview Press, 2005.

93. Lanquar, R. and R. Hollier., Le marketing touristique. Que sais-je? 1911. Paris, Presses Universitaires de France, 1986.

94. Gómez Martín, Ma., El Clima como activo del turismo: los folletos turı´sticos catalanes, El Territorio y su Imagen (vol. 1),

Malaga, Universidad de Malaga y Consejerıa de Medio Ambiente de la Junta de Andalucıa, 1999.

95. Gómez Martín, Ma., Climate potential and tourist demand in Catalonia (Spain) during the summer season. Climate Research. 32 (2006) 75.

96. Gómez Martín, Ma. and F. Lopez Palomeque, Tourism, territory and marginality: Principles and case studies. Paper presented at the Annual Conference of International Geographical Union Commission on Evolving Issues of Geographic Marginality in the Early 21st Century World. Stockholm, 2001.

97. Perry, A., Climate and weather information for the package holiday-maker. Weather. 48 (1993) 410. [98] Allen, P., Sunshine or your money back: French holiday companies take a gamble on the weather, 2009.

98. http://www.dailymail.co.uk/travel/article-1200739/Sunshine-money-French-holiday-companies-gamble-weather.html?ITO=1490(Accessed 9 July 2009).

99. Scott, D., Climate Change and Tourism in the Carolinas. Climate, Weather and Tourism Workshop. East Carolina University and NOAA Southeast Region. Grenville. 14-15 Nov, 2008. http://www.ecu.edu/cs-acad/sustainabletourism/upload/Daniel- Scott.pdf.

100. Smith, K., The effect of weather conditions on the public demand for meteorological information. Journal of Climatology. 1 (1981) 381.

101. Rutty, M. and D. Scott, Will the Mediterranean be 'Too Hot' for Tourism? In: Proceedings of 7th International Symposium on Tourism and Sustainability, Travel and Tourism in the Age of Climate Change. University of Brighton, Brighton, England, July 8-10, 2009.

102. Szalai, K. and T. Ratz, Tourist Perceptions of Uncertainty and Risk Associated with Extreme Weather Events. Paper presented at the 3rd International Conference on "Tourism

Future Trends", October 6, University Faculty of Tourism and Hotel Management, Egypt, 2006.

103. Kuehnel, J., personal communication (Director, JK Consulting and Enterprises, Toronto, Canada), 2009.

104. Trotz, N., personal communication (Science Advisor, Caribbean Climate Change Centre, Nassau, Bahamas), 2009.

105. Intergovernmental Panel on Climate Change (IPCC), Summary for Policymakers. In: Climate Change 2007: The Physical Science Basis. Contribution of Working Group I to the Fourth Assessment Report of the Intergovernmental Panel on Climate Change (S. Solomon, D. Qin, M. Manning, Z. Chen, M. Marquis, K.B. Averyt, M. Tignor and H.L. Miller, eds). Cambridge and New York, Cambridge University Press, 2007.

106. Ebner, D., Deep snow, deep pockets: It's a billion-dollar bet on climate change. Globe and Mail, 28 January, 2008.

107. Katzenberger, J., eds, Climate Change and Aspen: An Assessment of Impacts and Potential Responses, A report in the Aspen Global Change Institute Elements of Change series. AGCI, 2006.

108. Zeng, Z., Weather derivatives and weather insurance: Concept, application, and analysis. Bulletin of American Meteorological Society. 81 (2000) 2075.

109. Weather Risk Management Association, Weather risk market value plunges by 53 pct-survey, 2009. http://www.wrma.org/pdf/Weatherriskmarketvalueplungesby53pctsurveyReuters.com.pdf (Accessed 18 June 2009).

110. Weather Bill, Inc., Sunshine Guaranteed or Your Money Back: Breakthrough Weather Refund Promotions, WeatherBill White Paper. Weather Bill, Inc., 2008.

111. Speedwell Weather Derivatives, 2001: www.weatherderivs.com/press.html (Accessed 30 November 2001).

112. Chagnon, D., J. Sparks, A. Burgoyne and C. Hahn, Enhancing swimming pool management decisions with climate information. Meteorological Applications. 9 (2002) 461.

113. Farren, B. personal communication (Director of Grounds and Golf Course Management, Pinehurst Resort and Country Club, Greenville, North Carolina), 2008.

114. Government of Australia, Tourism and Climate Change – A Framework for Action. Department of Resources, Energy and Tourism, 2008. http://www.ret.gov.au/tourism/Documents/Tourism%20and%20Climate%20Change/climate_change_a_framework_for_action.pdf (Accessed 9 July 2009)

115. Curtis, S., J. Arrigo, P. Long and R. Covington, Climate, Weather and Tourism: Bridging Science and Practice. Centre for Sustainable Tourism, East Carolina University, 2009. http://www.research2.ecu.edu/Tourism/Documents/Summary%20Report.pdf.

116. McRitchie, P.. personal communication (Operator of McRitie Wine and Cider Works, Greenville, North Carolina, United States), 2008.

117. McCormick, C., personal communication (Executive Director, Outer Banks Tourism Authority, Greenville, North Carolina, United States), 2008.

118. Amelung, B. and D. Viner, Mediterranean tourism: Exploring the future with the tourism climate index. Journal of Sustainable Tourism. 14 (2006) 349.

119. Halifax Travel Insurance, Holiday 2030, 2006. http://www.hbosplc.com/media/pressreleases/articles/halifax/2006-09-01- 05.asp?section=Halifax.

120. Osborne, H., Climate change put ski resorts on slippery slope. Guardian.co.uk, 14 February, 2007. [121] Munns, R., Climate change to reduce European holidays to the Mediterranean, 2008.

121. www.content4reprent.com/environment/climate-change-to-reduce-european-holidays-to-mediterranean. (Accessed 28 January 2009).

122. The Independent, France: tuned in to the long-range forecast, 30 January, 2008. [123] Couttswoman, Where to buy ski property to avoid global warming, 2008.

123. www.couttswomand.com /2008/february/travel/where-to-buy-ski-property-to-avoid-global-warming (Accessed 5 February 2008).

124. Stewart, T.R., R. Pielke Jr and R. Nath, Understanding user decision-making and the value of improved precipitation forecasts: Lessons from a case study. American Meteorological Society, February, 2004, 223.

125. Keltie, D. Ski Operations Manager's Decision-making Under Uncertainty. Unpublished Master's Thesis, Department of Recreation and Leisure, University of Waterloo, Canada, 2007.

126. Mieczkowski, Z. The tourism climate index: A method for evaluating world climates for tourism. The Canadian Geographer. 29 (1985) 220.

127. de Freitas, C.R., D. Scott and G. McBoyle, A second generation climate index for tourism (CIT): Specification and verification. International Journal of Biometeorology. 52 (2008) 399.

128. Scott, D., C. de Freitas and A. Matzarakis, Adaptation in the Tourism and Recreation Sector. In: Biometeorology for Adaptation to Climate Variability and Change (K. Ebi, I. Burton and G. McGregor, eds), 2008.

129. Barnes, B.The heat wave vacation. The Wall Street Journal. 2 August, 2002.

130. Scott, D., G. McBoyle and B. Mills, Climate change and the skiing industry in Southern Ontario (Canada): Exploring the

importance of snowmaking as a technical adaptation. Climate Research. 23 (2003) 171.

131. Scott, D. G. McBoyle and M. Schwartzentruber, Climate change and the distribution of climatic resources for tourism in North America. Climate Research. 27 (2004) 105.

132. Scott, D.. B. Jones and G. McBoyle, Climate, Tourism and Recreation: A Bibliography 1936 to 2005. University of Waterloo, Department of Geography, Waterloo, Ontario, 2006. http://www.fes. uwaterloo.ca/geography/faculty/dscott.html.

133. Lise, W. and R. Tol, Impact of climate on tourist demand. Climatic Change. 55 (2002) 429.

134. United Nations World Tourism Organization (UNWTO), Recommendations on Tourism Statistics, Madrid, UNWTO,1994.

135. Lennon, J. Tourism Statistics: International Perspectives and Current Issues, New York, International Thomson Business Press, 2001.

136. Katz, R.W. and A.H. Murphy, Economic value of weather and climate forecasts. Climatic Change. 45 (2000) 601.

137. National Oceanic and Atmospheric Administration (NOAA), Geostationary Operational Environmental Satellite system (GOES) GOES-R sounder and imager cost/benefit analysis (CBA). NOAA NESDIS Office of Systems Development: Silver Spring, 2002.http://www.economics.noaa. gov/?file=bibliography#noaa.2004 (Accessed 9 July 2009).

138. National Oceanic and Atmospheric Administration (NOAA), Geostationary Operational Environmental Satellite system (GOES) GOES-R sounder and imager cost/benefit analysis (CBA) - Phase III. NOAA/NESDIS Office of Systems Development, Silver Spring, 2004. http://www.economics.noaa.gov/?file=bibliography #noaa.2004 (Accessed 9 July 2009)

139. Centrec Consulting Group, LLC., An Investigation of the Economic and Social Value of Selected NOAA Data and Products for Geostationary Operational Environmental Satellites (GOES), A report submitted to NOAA's National Climatic Data Center. Centrec Consulting Group, Savoy, 2007. http://www.economics.noaa.gov/?file=bibliography#noaa.2004 (Accessed 9 July 2009).

140. National Research Council, Critical issues in weather modification research. Committee on the Status and Future Directions in US Weather Modification Research and Operations, National Research Council. Washington, 2003.

141. Drabek, T.E., Disaster Evacuation and the Tourist Industry. Institute of Behavioral Science, Program on Environment and Behavior, Boulder, University of Colorado, 1994.

142. Drabek, T.E. Disaster Evacuation Behavior: Tourists and other Transients. Institute of Behavioral Science, Program on Environment and Behavior. Boulder, University of Colorado, 1996.

143. Drabek, T.E. Disaster evacuations: Tourist-business managers rarely act as customers expect. Cornell Hotel and Restaurant Administration Quarterly. 41 (2000) 48.

144. Sorensen, J. Hazard warning systems: Review of 20 years of progress. Natural Hazards Review. May 2000, 119.

145. Scott, D. B. Jones and H. Abi Khaled, Climate Change: A Long-Term Strategic Issue for the NCC. Implications for Recreation-Tourism Business Lines. Report prepared for the National Capital Commission, University of Waterloo, Department of Geography, Waterloo, Ontario, 2005b.

146. World Meteorological Organization (WMO), World Climate News, 27th edition. Special Issue on Climate and Tourism, 2005b. http://www.wmo.ch/web/catalogue/New%20HTML/frame/engfil/wcn/wcn27.pdf. (Accessed 5 June 2006).

147. KPMG, Climate Changes Your Business, 2008. www.kpmg.nl/sustainability.

148. Hall, C.M. Tourism and climate change: Knowledge gaps and issues. Tourism Recreation Research. 33 (2008) 339.

Chapter 7

Forms of Pilgrimage at the Shrine of Khāled Nabi, Northeastern Iran

Mehdi Ebadi

Department of Geography University of Heidelberg

ABSTRACT

Pilgrimage (religious tourism) is one of the fastest growing forms of tourism. Nevertheless, there is still a gap between abstract theory and empirical research about this form of tourism in the literature. The dearth of studies is even more glaring in the field of *Ziyārat* or pilgrimage in Islam which in spite of its importance and wide extended practice has been mostly ignored in tourism and geographic literature. The present study features one such example that is (almost) unknown within the community of tourism and geography researchers. In Iran, religious pilgrimage has a long tradition. Numerous sacred places with various rituals and traditions, practiced by pilgrims all around the country, indicate its antiquity (before the Islamic periods). Among the most practiced forms of religious pilgrimage are visits to several thousand shrines, which are known in Iran as *Ziyāratgah* (lit. place of visit) or *Imāmzādeh*. One of these *Ziyāratgah* is the pre-Islamic shrine of Khāled Nabi (also known as *Halat Nabi*), linked to a legendary Christian holy man from the 6th century AD. The shrine lies in the northeastern Iranian province of Golestan, in a region called Turkmen-Sahra. Despite its relatively remote setting, every year more than 90,000 travellers visit this shrine. This paper considers the variety of pilgrimage forms at the shrine and shows that the travellers to Khāled Nabi shrine are not homogenous and comprise different types of visitors. In addition to secular motivations, based on the visitors' curiosity, three forms of pilgrimage, namely, 'religious pilgrims', 'cultural pilgrims' and 'nostalgic pilgrims' have been identified.

Keywords: forms of pilgrimage, ziyārat, imamzadeh, Khāled Nabi, Turkmen Sahra, Iran

INTRODUCTION

Worldwide, national and international religious and spiritual travel in the form of pilgrimage and religious tourism is one of the fastest growing forms of tourism (WTO, 2011; Olsen and Timothy, 2006; Rinschede and Bhardwaj, 1988). Since the early days of human history a multitude of sacred places around the world like shrines, temples, churches, landscape features, religious festivals and so on, have prompted hundreds of millions of pilgrims to visit. Worldwide, numerous cities have been founded and developed because of pilgrimage and religious tourism. In spite of huge changes in modern society, religious motivated travels are still an essential fixture in various religious cultures and are even finding a renaissance in the world and Europe (Sommer, 2012). Accordingly, in Europe alone, there are more than 6000 Christian pilgrimage sites which in total attract 60 to 70 million pilgrims every year (Antz, 2012).

Despite its importance, literature comparing studies about religious pilgrimage to other forms of tourism, and its connection to recreational (secular) tourism, have been relatively overlooked. This lack of pilgrimage studies in tourism and geographic literature is even more extreme in relation to *Ziyārat* or shrine pilgrimage in the Islamic contexts, in spite of its significance and extensive practice which have been mostly ignored.

In contrast to the Hajj (pilgrimage to Mecca) which is considered as the 'official' or 'normative' form of pilgrimage in Islam, *Ziyārat* have been considered as the 'popular' (i.e. *völkisch*) form of pilgrimage, which is full of local cultural aspects and practiced in different forms among people from different cultures. Therefore, there is a close connection between *Ziyārat* and (cultural) tourism.

In Iran, religious pilgrimage (*Ziyārat*) has a long tradition. Numerous often pre-Islamic sacred places in the country indicate its antiquity and its importance in the life of Jewish, Christian, Zoroastrian and Moslem Iranians. In recent decades, especially after the 1979 revolution, the importance of pilgrimage found its pinnacle, since people have been encouraged to do pilgrimage to holy shrines. Accordingly, several shrines and holy places have been restored or developed by the government and people. Some of these holy places, in addition to attracting varieties of pilgrims, also attract cultural tourists on account of their antiquity. Among the most practiced forms of religious pilgrimage are visits to several thousand Islamic shrines, which are known in Iran as *Ziyāratgah* (lit. place of visit) or *Imāmzādeh*.

In addition to important Shia' shrines in the cities of Qom, Mashhad and Shiraz, there are some shrines in Iran which are overwhelmingly visited by Sunni Moslems. The shrine of Khāled Nabi is one of those which, during the past three decades has developed from an unknown isolated shrine to a weekly venue of assembly, especially for the Turkmen community which is one of Iran's lesser known minorities. According to the available literature, the shrine belongs to a putative Christian missionary from the 6th Century. The popularity of Khāled Nabi among the Turkmens has attracted the attention of curious Iranians (domestic and from abroad) and a small number of foreign tourists. Accordingly, despite its relatively remote setting, every year more than 90,000 travellers visit this site.

The main objective of this article is to consider the different tourism functions of the site of Khāled Nabi, as one of the most popular historical shrine in the north east of Iran. Accordingly, in the theoretical part, after reviewing the available literature about tourism and pilgrimage, their connections will be discussed. We seek to shed more light on the possibilities of categorizing visitors to Khāled Nabi shrine, according to their motivation.

Research Methodology

Because of the multidisciplinary characters of pilgrimage and tourism (as noted by Bieger, 2006 also by Schmude and Namberger, 2010), for the collection of data for this research, the author has relied on several traditional qualitative and qualitative methods, notably general observation, participant observation, in-depth interviews with key persons and distributing questionnaires among the visitors. To provide insight into the general opinions of visitors, a total of 180 questionnaires that contained open-ended and closed questions were distributed among Turkmen and non-Turkmen pilgrims/visitors in the age group of 18-80, of which 100 were completed.[2] Data were coded and entered into SPSS. Field work was carried out from April to June 2011 and also in August 2011. Since the number of visitors vary throughout the week, to obtain the best results, the investigation was conducted across a whole week to include periods with maximum, average, and minimum number of visitors.

FROM RELIGIOUS PILGRIMAGE TO SECULAR TOURISM

The definition of pilgrimage and tourism and their connections have been discussed extensively in the tourism literature.

The term 'tourism' is considered to be a challenging multidisciplinary and interdisciplinary field of study by most accounts. Researchers working in disciplines as varied as geography, anthropology, economics, marketing, sociology, history, theology, psychology, political studies, law, environment and so on, have tried to define and explain the precise nature of this activity. One may subscribe to the observation of two prominent experts in the field of tourism research, (Smith, 2001; Murphy, 1985) who indicate: 'there are as many definitions of tourism as there are disciplines and investigations' (Smith, 2001, p.15) and 'there are as many types of tourist as there are motives for travel' (Murphy, 1985, p.5).

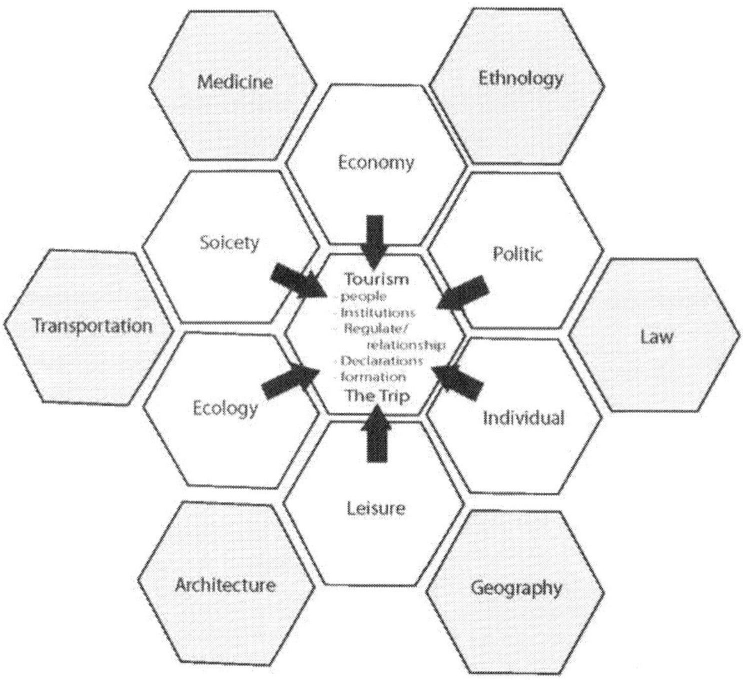

Figure 1 : Holistic Model of Tourism

Source: Freyer (2011:47) original in German, translated by the author

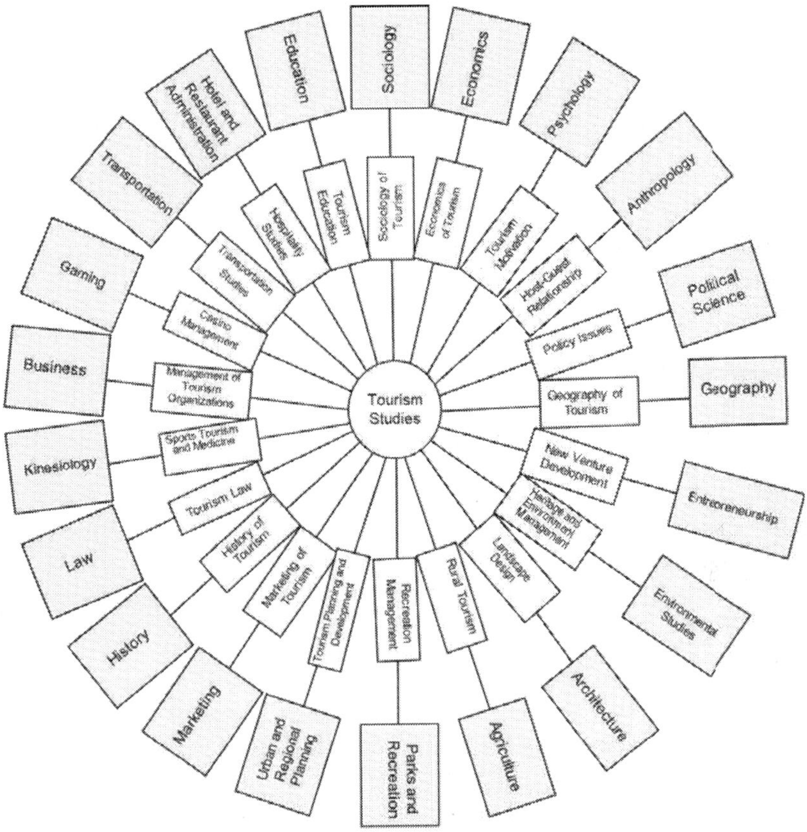

Figure 2 : Model of Knowledge Productions in Tourism

Source: Originally proposed by Jafari (1981) cited in Goeldner & Ritchie, 2009

The difficulty of finding the appropriate and all- encompassing definition for 'tourism' is certainly reflected in the many, sometimes contradictory opinions within the scholarly community on this often, poorly understood phenomenon. According to Williams (2009), the problem of defining tourism arises, since a single term should designate varieties of meanings [interpretations] which include an area of study in a range of disciplines. Some scholars like Freyer (2011) and Jafari (1981), have tried to define tourism by introducing holistic models which indicate the area study of tourism, Figures 1 & 2.

Like tourism, the multifaceted phenomenon of pilgrimage is frequently mentioned in the sphere of travel. Its interdisciplinary character has engendered a lively debate among scholars, who have used several approaches to capture *pilgrimage* in a few clear-cut and generally valid definitions. It is agreed generally that 'traditional' pilgrimage is a journey inspired by spiritual or religious reasons to a 'hallowed place' that could consist of:

a. Elements of the natural and geographical environment such as mountains, rivers, springs, trees, caves, lakes, islands, groves or even animals.

b. Religious edifices such as churches, cathedrals, mosques, sites of revelations or the activities of the religious founders, temples, synagogue or shrines.

c. General venue for religious and spiritual activities, ritual or festivals (Cohen, 2000:438, 2006; Hosta and Limon, 2009; Davidson *et al.*, 1990; Morinis, 1992 and Shackley, 2006)

Pilgrimage attractions comprise numerous forms of sites worldwide; ranging from small regional shrines to world-famous sanctuaries these attract millions of believers. The prominent German geographer, Rinschede (2000:37) describes religious attractions as follows:

[These] are places or events which are recognized by individuals or groups as worthy of devotion and visitation. As religious or sacred phenomena, they stand out from the profane and commonplace. They can be distinguished primarily by the intensity of the perceived holiness.

FORMS OF PILGRIMAGE

Contrary to the commonly held impression, not all visitors to religious sites are purely religious pilgrims - people with varieties of motives visit religious sites. In fact, pilgrimage is an activity which cuts across many human cultures and traditions. Several scholars have attempted to capture different features of religious travel. In their research about the motivations of Polish pilgrims, Jackwoski and Smith (1992:93) suggest that although 'a pilgrimage has traditionally been described as a religiously motivated journey to a sacred place', it is actually possible to distinguish two forms of this, viz. a pilgrimage motivated (purely) by religion (or 'true pilgrimage' as they called it), and a pilgrimage motivated by knowledge. In the former category, the pilgrims, who are mostly poorly informed or blissfully unaware of the historical or cultural significance of the place, generally spend their time in meditation, prayer and performing religious rites. The latter category, the so-called *knowledge*-based pilgrims (or, according to the definition of the authors, 'religious tourists'), includes those pilgrims whose motivation for the journey is largely 'knowledge-based' and their primary pursuit is to have 'an experience with the [pilgrimage] route', on the basis of the information they have gathered from their surroundings, the countryside through which they pass and the people they meet.

Generally, as Eade and Sallnow (1991:3) have noted: *pilgrimage as an institution cannot actually be understood as a universal or homogeneous phenomenon but should instead be deconstructed into historically and culturally specific instances.*

It appears therefore that it is possible to find a huge diversity in the field of pilgrimage. In fact, the concept of pilgrimage exists in all major religions but with various meanings depending on the religious structure. In other words, people from different cultural, geographical and socio- political milieus and background might have different perceptions of the meaning surrounding pilgrimage, for example in relation to issues such as *holiness*. A good example for this is the visits of Iranian people to the mausoleum of the famous Persian poet Hafez located in Shiraz (in the south of Iran). For Iranians, who have learned many of Hafez's ghazals by heart and have been immersed in their mystical meaning and spirit, such a visit acquires a very spiritual dimension that borders on a religious pilgrimage. At the same time, for many foreign tourists, who are not deeply familiar with Hafez's poetry, the journey is a mere stop-over in their itinerary and the tomb happens to be one of the more important attractions of Shiraz.

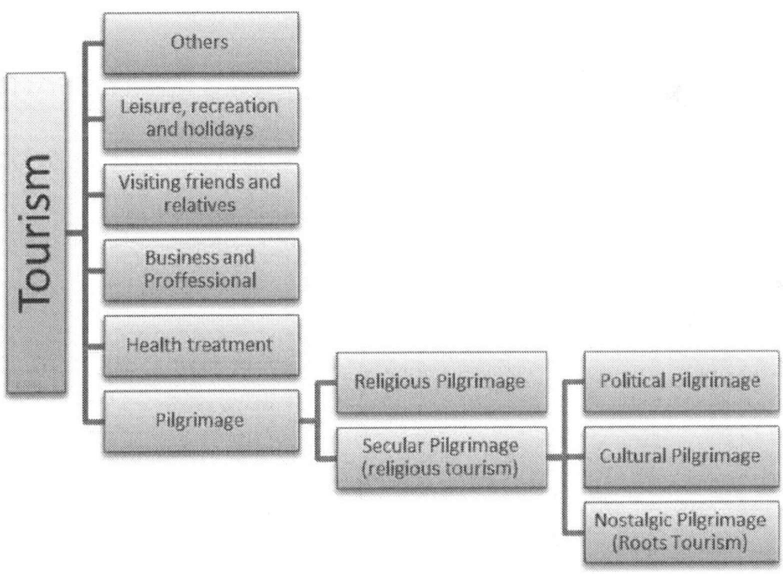

Figure 3 : Forms of Pilgrimage in Tourism Studies

As indicated, the reasons to be on a pilgrimage are manifold. Although, the key motivation in pilgrimage is religion, in many cases other reasons which are similar to those for (secular) tourism also accompany it. In addition, not all religious phenomena that are venerated or visited by devout believers or tourists have equal status or perceived holiness. As pointed out by Blackwell (2009:33):

motivation, furthermore, can change when the individual switches activities, for example, from being a pilgrim to a tourist and vice-versa, often without the individual being aware of the change.

PILGRIMAGE VS. RELIGIOUS TOURISM

Because of the diversity of motivations involved, like tourism, the term 'pilgrimage' is still a general term which has been widely used both in sacred and secular contexts, therefore, based on the tourism definition of the UNWTO[3], pilgrimage can be classified as a subcategory of 'tourism'. Reviewing available pilgrimage literature[4] indicates the use of the term *religious tourism* (instead of pilgrimage) by many scholars which has caused several ambiguities. Firstly, the term *tourism*, in most of these publications refers to recreational secular (non-religious) forms of activity, while in other important texts (such as the Encyclopaedia of Tourism or UNWTO's definition) tourism has been used as a general term which includes both religious and non- religious motivated travel. Secondly, 'religious tourism', being applied to both religious and semi- religious visitors is still a broad term. Due to the dearth of appropriate terms[5], we shall classify the nature of the visits to sacred places (depending on the motivations of the visitors), in two main categories: 'religious' and 'secular' pilgrimage, see Figure 3.

Religious Pilgrimage

According to Figure 3, *religious pilgrimage* (also known as 'sacred' pilgrimage) is the pilgrimage performed by traditional believers who are concerned only with holiness and worship in reference to a higher or supernatural deity or being (s). 'Mystico-religious sites' are places which are mainly the destination of 'religious pilgrims.' In fact as Rinschede (2000:37) states:

[These sites] *are perceived* [by them] *as the most sacred, because they believe that God and man are brought into direct context through them.*

Typical examples include the pilgrimage undertaken by followers of Catholic Christianity to Lourdes or Fatima, the *hajj* to Mecca by Muslims and Hindu pilgrimage to the River Ganges. The expectations for such visits are usually to gain a blessing (or *baraka* in Arabic), a cure of illness, ailment, etc.

Secular Pilgrimage

Secular pilgrimage which is mostly classified by scholars as *religious tourism,* includes that group of visitors (pilgrims) who visit mostly sacred places and are motivated by religion *in combination* with other kinds of motivations, namely quasi-spiritual motivations. In fact, the changes in life style and strong secularization of modern life encourages many 'modern' pilgrims to travel for reasons rather than religious, like cultural, political, knowledge-seeking or simply a quest for one's roots and identity (religious, cultural, political or familial.) Secular pilgrimage in other words, fulfils both religious and recreational needs of the travellers.

Grasping the exact nature of 'secular pilgrimage' (religious tourism) has raised a series of debates among the scholars. Like other forms of tourism, drawing an exact line between different types of pilgrimage seems nearly impossible. However, we shall classify secular pilgrimage into three forms: cultural; political and; nostalgic pilgrimage.

Cultural Pilgrimage: There are many ancient religious and spiritual sites around the world that attract many visitors not because of their religious or sacred nature (in the traditional sense), but because of their remarkable cultural and historical heritage magnetisms. In other words, these places, which are mostly enlisted in UNESCO's List of World Heritage Sites, are known by their visitors due to their admirable architecture, appearance or historical importance, rather than their religious meanings. Sites include places such as shrines, synagogues, pagodas, monasteries, temples, caves, mountains, groves, sites associated with (persecuted) religious minorities or religious deviance, mosques, stupas, ancestral halls, pyramids, ashrams, lakes, islands, cemeteries and other burial places, towns and sites of violent conflicts, including battlefields (Stausberg, 2011; Sharpley, 2012).

Because of the close relationship between pilgrimage and cultural events, we label this group of travellers as 'cultural pilgrims'. Because of the close connection between 'culture' and 'religion' in pilgrimage, some scholars (like Digance, 2003) view pilgrimage as a 'cultural phenomenon' which historically connects people from diverse cultures and ethnic groups. In fact, cultural pilgrimage deals with the visits of cultural tourists, mostly out of curiosity about different cultures and religions, travel to religio-cultural festivals, historical sacred places or cultural figures. Accordingly, sites of religion worldwide can be assumed as sites of heritage. Da Graca Mouga Pocas Santos (2003: 38) describes these sort of pilgrims as those who might be interested in the 'meaning and symbolism of religious places, monuments and practices, and seeks to deepen [their] knowledge [about] these things.' She conceived that these visitors don't achieve the status of *pilgrim* or even *religious tourist*, since:

1) These visitors don't have any sense of the spirituality of these places,

2) They don't actively participate in religious acts which take place in these places.

Indeed, cultural pilgrims are mostly defined due to their knowledge-based motives (such as learning about the history and culture of a religious place) rather than purely religious or pleasure-seeking motivations.

Generally, because of the complexities of these motivations, in some cases, a religious site might fulfil numerous functions. But these roles can not only be determined from outside and they depend on the needs and purposes of the visitors. Accordingly, for example, a religious building can simultaneously perform two functions. It can firstly, satisfy the religious needs of the visitor (i.e. Gottesdienst) and secondly, it might comprise of a special attraction – like its unique architecture or wall paintings (Sommer, 2012)

Among the most popular cultural pilgrimage sites are the Notre Dame Cathedral in Paris, Cologne Cathedral in Germany and many religious temples in Asia, which are visited by International tourists. Other sites that function as places of cultural pilgrimage are those related to famous literary figures such as the Persian and British poets Rumi and Shakespeare or artists like the American singer Elvis Presley, which elicit adoring sort of feelings during the visits of their adherents and disciples.

In addition, religious festivals might also classify as cultural pilgrimage. These feasts which have both cultural and religious rituals and traditions are visited by both secular and religious tourists.

Political Pilgrimage: 'Political pilgrimage' as another form of secular pilgrimage, is related to travel (mostly motivated by governments) of people to countries, who wish to visit quasi- religious attractions and ceremonies (mostly funerals) related to national leaders, heroes or war victims or similar in order to pay respect - merely out of curiosity. People with different motivations (from strong emotion to simple curiosity) visit these places. Among popular political pilgrimage sites are the intact body of Lenin in Lenin's Mausoleum in Moscow; Nelson Mandela's prison on Robben Island, or the *ārāmgah* (resting place) of Imām Khomeini near Tehran.

Nostalgic Pilgrimage: Finally, 'nostalgic pilgrimage', also known as 'roots tourism' is another form of secular pilgrimage, in which the 'pilgrim' 'quests' to know more about their social, religious, cultural, national or ethnic identity, and thus, visit places which are presumed to have relations to their ancestors or their past. The search for identity is even stronger among the religio-ethnic minorities. Basu (2004:173) defines nostalgic pilgrimage (which he called 'roots- tourism') as a metaphorical and terrestrial journey to home (homecoming), to the source and cradle of belonging. According to Sharpley and Sundaram (2005), nostalgic pilgrims are usually strongly motivated to connect with their history.

Photo 1 : The shrine of Chupān Atā and *Kopet-Dāgh* mountains ('thousand hills') late summer 2011

Photo 2 : Historical graveyards near Khāled Nabi, 20.08.2011

Regarding the motivations of 'on foot' pilgrims in Europe, Brämer (2012:57) links this to some sort of nostalgic motivations such as a 'search for old values, simple life and partially seek for Christian symbols'. On occasions, this question of identity is more existential and philosophical, and inspires travellers to visit for relatively long periods of time, for what Basu sees as 'a quasi-mystical finding of oneself in others and others in oneself' (2004:168).

FORMS OF PILGRIMAGE AT KHĀLED NABI SHRINE

Khāled Nabi Shrine – known also as *imāmzādeh-e Khāled Nabi* - is one of the most sacred and popular sacred shrines for Iranian Turkmens. This shrine complex is located on the southern flank of 712-m.-high Mt. *Gockche Dāgh* in the district of Maraveh Teppeh of the Turkmen Sahra. The summit is called *tangry daq* (Mount of God in Turkish) by Turkmen. *Gockche Dāgh* is surrounded by the low, hearther-clad hills of Kope Dāgh, which are referred to by locals as the *hezar tappeh* (thousands hills). This stretch of hills are located between Iran and Turkmenistan (*Photo 1*). The area of Khāled Nabi consists of three major shrines of

- Khalid ibn al-Sinan (simply known by locals as Khāled Nabi),
- *Chupān Atā* and
- *Alim Bābā*.

About 3 Km from the shrine of Khāled Nabi, a former burial mound is located (Photo 2). This graveyard is strewn with mostly phallic shaped gravestones, which appear to be of great (pre or early Islamic) antiquity (Stronach and Royce, 1981:147).

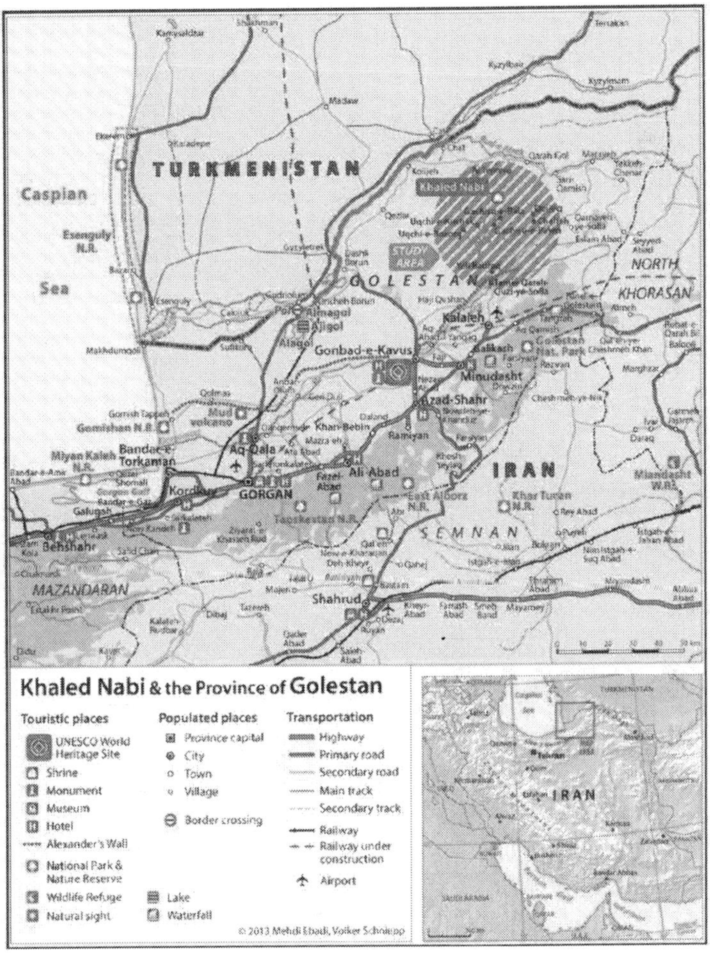

Map1 : Khāled Nabi & the provience of Golestan

The history of pilgrimage to Khāled Nabi is generally unknown. It is unclear when the site began to act as a religious shrine especially for local Turkmens. There is no information about the time of construction of the shrines at the complex. As part of a PhD research project carried out by the author, it emerged that during the last 30 years, Khāled Nabi has been developed enormously and has started to generate public interest - establishing itself as an important regional shrine for Turkmens. Reviewing the demographic aspects of the visitors indicates that the visitors are mainly male (85%); include more Sunni Turkmen (66%) than Shia non-Turkmen (34%); and are relatively young (65% of visitors are under 40 year old). Furthermore, the main purpose of 32% of visitors to the Khāled Nabi shrine is pilgrimage. A further 26% have recreational (non-religious) motivations and the remaining 42% are interested in both religious and recreational reasons (Figure 4).

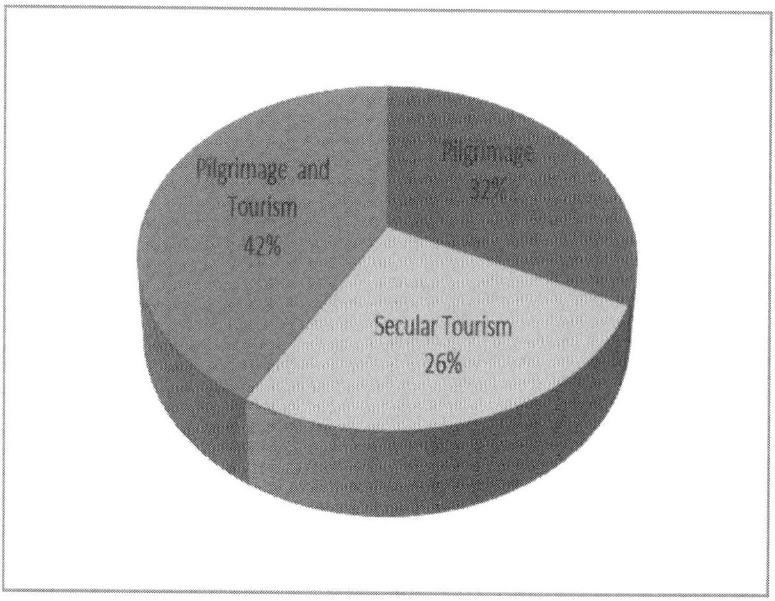

Figure 4 : Form of activities among all the visitors

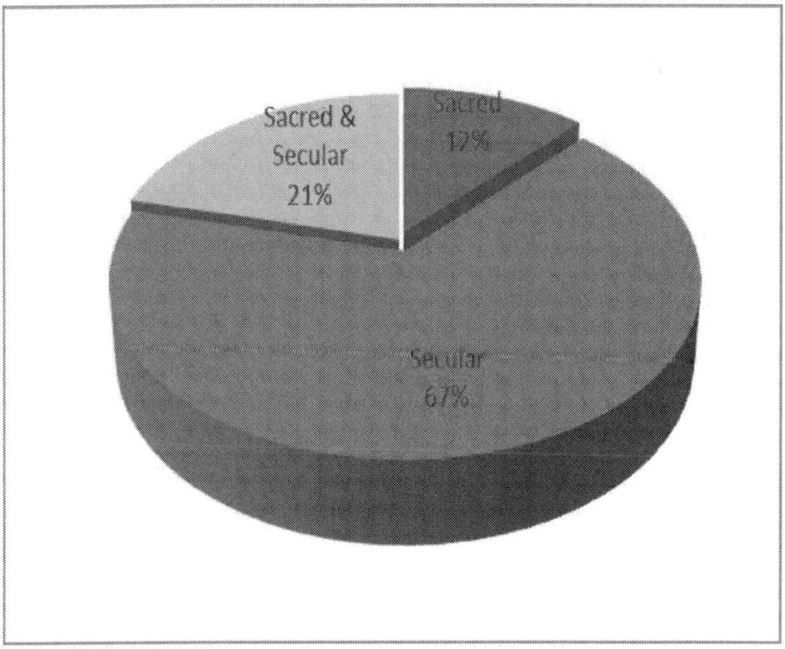

Figure 5 : The primary purposes of visits among non-

Interestingly, the motivations for visit are different for Turkmens and non-Turkmens. As can be seen in Figures 5 & 6, 42% of Turkmens visited Khāled Nabi just for 'sacred' (pilgrimage) reasons and the majority of the remaining visitors (52%) visited for both 'sacred' and 'secular' motivations. Only 6% of Turkmens visited the site as 'secular' (non- religious) tourists.

In contrast, a minority (12%) of non-Turkmens visited the shrine for 'sacred' reasons, with 21% visiting for both 'sacred' and 'secular' motivations. But overall a clear majority (67%) visited Khāled Nabi for 'secular' (non-religious) reasons, see

In addition to its religious appeal, the sanctuary of Khāled Nabi offers several other forms of entertainment that are a draw for visitors with non -religious motivations. It is important to consider that these motivations are changing.

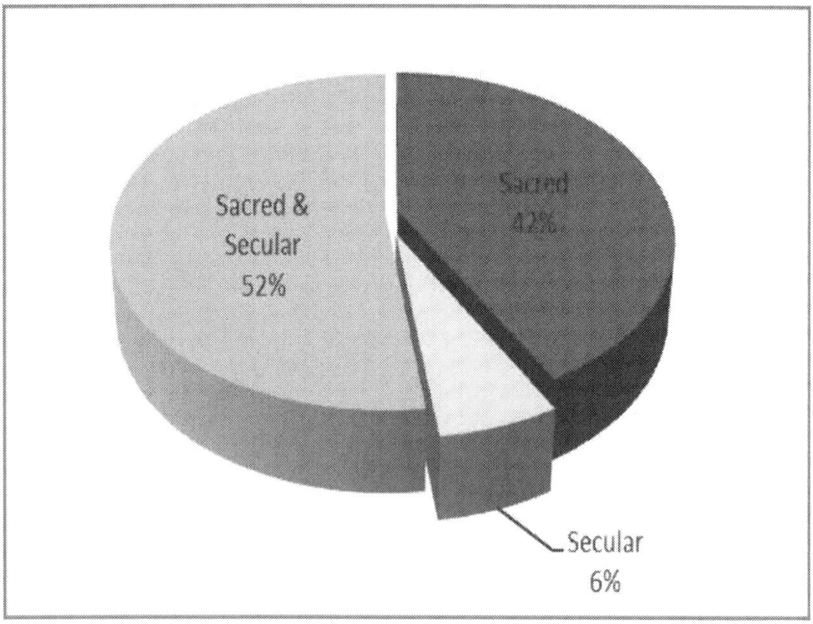

Figure 6 : The primary purposes of visits among Turkmen travellers

Often, visitors may seek multiple experiences even on short trips and switch their motivations unconsciously as they visit a destination.

THE SACRED-SECULAR AXIS

In this research, to solve this problem of *simultaneous* and *overlapping* motivations, the classification of visitors is based on their *self- described* motivations derived from the questionnaire (Figure 7). Based on the visitor's motivation three forms of pilgrimage: 'religious', 'cultural' and, 'nostalgic' have been recognized at the site.

Religious Pilgrims:

These visitors are solely preoccupied with the holiness of Khāled Nabi. They visit the shrine out of religious belief and their main activity is praying. They come to Khāled Nabi specifically to ask for help and assistance in problems related to personal finance, health, and fertility or finding a suitable partner (for themselves or for their

children). The believers who have such a vow, tie coloured rags on the branches of a tree next to the shrine in order to have their wishes granted. Those who have their wish granted donate a small carpet to the shrine or sacrifice a lamb or calf, depending on the nature of their wish.

Cultural Pilgrims

This category includes non-Turkmen visitors who believe in the holiness of the site, but unlike their Turkmen fellows, do not come to the site to make a vow; they are more interested in the spiritual atmosphere of the sanctuary. These visitors also show interest in the natural and historical attractions of the site.

Nostalgic Pilgrims

The last category includes 'nostalgic pilgrims' who are mostly motivated by religio-cultural reasons. They consist of young and middle-aged Turkmens in a middle or high income group, who, instead of being driven by (purely, mainly) religious motivations, are mostly interested in the historical and cultural aspects of the site. Among their motivations for visit, we encounter some concepts like:

- a visit to confirm their roots,
- satisfying nostalgic feelings about the good past- days,
- quest of identity,
- educational reasons,
- family togetherness or similar.

These Turkmen visitors, who usually have an academic educational background, visit Khāled Nabi in search of their cultural and religious identity. Simultaneously, they might be interested in nature as well. Because of the difficulties experienced by Turkmen in modern Iranian society, Khāled Nabi offers a chance for cultural unity through pilgrimage or the simple act of gathering together (without being under the constant scrutiny of the local authorities or other Iranian citizens). In fact, this category have transferred and copied their own central Asian ancestral values and customs to the site and have come to accept Khāled Nabi as their own original sacred place. According to pilgrimage classification, we shall classify them as nostalgic pilgrims.

Figure 7 The Sacred-Secular Axis of Different Forms of Activities at Khāled Nabi Shrine

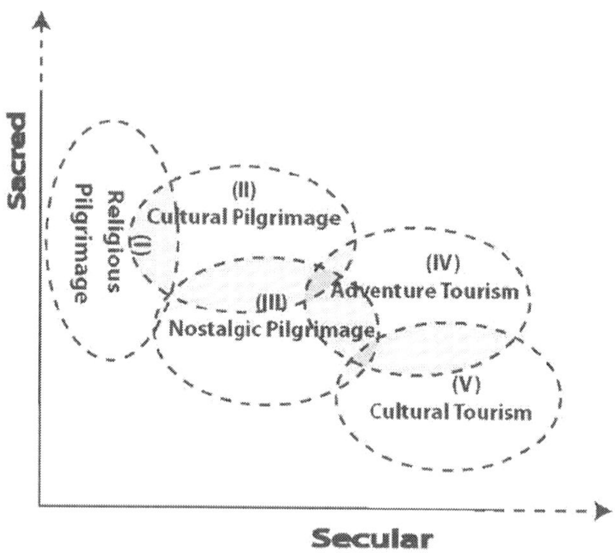

Source: Author

Figure 7 The Sacred-Secular Axis of Different Forms of Activities at Khāled Nabi Shrine

CONCLUSION

Reviewing literature, nearly all scholars acknowledge the close relationship between leisure tourism and pilgrimage and admit that their connections have been blurred. In fact, religious sites, in addition to satisfying the religious needs of their religious pilgrims, provide various forms of leisure / entertainment for tourists.

Among more than 100 *Imāmzādeh* which are located in the Turkmen Sahara (*northeastern Iran),* the shrine of Khāled Nabi is the most well-known one, especially during the last three decades as it has developed from an unknown, remote place to the most important gathering centre for Turkmen. The shrine derives from a putative Christian Missioner from the 6th Century. Located in the Perso-Turkic cultural boarder line of Iran and Central Asia, the rituals that are held by Turkmen in Khāled Nabi are fairly unique in Iran and cannot obviously be connected with mainstream (Shia) Islamic practice.

The main findings of the current study included the investigation of different available forms of tourism and pilgrimage at Khāled Nabi shrine. Accordingly, in order to explain the *tourist* and/or *pilgrim* role(s) of the visitors and the possibility of classifying them, their demographic features (age, gender, social and economic situation), perceptions, motivations and behaviour were considered.

The results of the study show that, contrary to the prevailing view, visitors to the shrine of Khāled Nabi are not homogenous and comprise of different types of visitors. The geographical backdrop of the shrine with its beautiful natural scenery (especially in spring), an ancient burial ground, and the ethnic allure for the Turkmens, attracts a range of different visitors with different motivations. They come with religious motivation (32%), secular (recreational) motivation (26%) and semi religious incentives - both religious and recreational (42% of all visitors).

Generally, the visitors to Khāled Nabi can be classified into two common groups: Turkmen and non-Turkmen visitors. The Turkmen visitors who constitute the majority (65%) are mainly (42%) interested in religious pilgrimage or combine pilgrimage with other forms of tourism (52%). During the Islamic holidays, hundreds of Turkmen pilgrims arrive at Khāled Nabi to pray. Nonetheless, for some other Turkmens (4%), Khlaed Nabi is just a place for meeting other fellow Turkmens, having picnics and spending a day with their families. Because of these characteristics, different rituals and activities in Khāled Nabi can be observed.

In contrast, the majority (58%) of the non- Turkmen travellers are curious Iranians (domestic and from abroad) and a few foreign tourists, who visit Khāled Nabi for secular reasons. Only 12% of the non-Turkmen visitors visit the shrine as religious pilgrims.

Based on the visitor's motivation three forms of pilgrimage: 'religious', 'cultural' and, 'nostalgic' have been recognized at the site. Religious pilgrims includes those visitors (mostly Turkmens) who out of religious belief visit the shrine and their main activity is praying. Cultural pilgrimage includes non-

Turkmen visitors who believe in the holiness of the site (figures 6), unlike the Turkmen religious pilgrims, these non-Turkmen visitors do not come to the site to perform a religious ritual (e.g. making a vow). Nostalgic pilgrimage includes young and middle age Turkmens, who, instead of purely motivated by religion are interested in the historical and cultural aspects of the site. Among their motivations to visit, we encounter concepts such as: a visit to confirm their roots, satisfying nostalgic feeling about the good- past-days, quest of identity, educational reasons, family togetherness, or similar.

REFERENCES

1. Antz C (2012) Im guten Glauben – Spritualler Tourismus als Reisemarkt der Zukunft? [In good faith – spiritual tourism as future travel market?]. In: Hopfinger H, Pechlaner H, Schön S and Antz C (eds.) *Kulturfaktor Spritualität und Tourismus*, Göttingen: ESV, 225-252.
2. Ayaz M (2012) Between history and memory, A case study of a martyr mausoleum in Iran. In: Khosronejad P (ed.) *Saints and their Pilgrims in Iran and Neighbouring Countries*: Lighting Source, 83-105.
3. Basu P (2004) Route metaphors of 'roots-tourism' in the Scottish Highland diaspora. In: Coleman S and Eade J (ed.) *Reframing Pilgrimage*: Routledge Chapman & Hall, 153-178.
4. Bieger T (2006) *Tourismuslehre-Ein Grundriss* [*Tourism – A teaching plan*]: UTB.
5. Blackwell R (2009) Motivation for pilgrimage: using theory to explore motivations. In: Ahlbäck T and Dahla B (ed.) *Pilgrimages Today: based on papers read at the symposium on pilgrimages today held at Åbo*, Finland, on 19-21 August 2009, 24-37.
6. Brämer R (2012) Heile Welt zu Fuß. Empirische Befunde zum spirituellen Charakter von Pilgern und Wandern [Heal the world on foot. Empirical findings on the spiritual character of pilgrimage and hiking]. In: Hopfinger H, Pechlaner H, Schön S and Antz C (eds.) *Kulturfaktor Spritualität und Tourismus*, Göttingen: ESV, 47-58.
7. Cultural Handicraft Tourism News, CHTN (2013) Last year more than fifty million travelers visit the shrines in Iran. Available at www.chtn.ir
8. Cohen E (2000) Pilgrim. In: Jafari J (eds.) *Encyclopedia of Tourism*: Routledge, 438.

9. Cohen E (2006) Religious tourism as an educational experience. In: Timothy DJ and Olsen DH (eds.) *Tourism, Religious and Spritual Journeys*: Routledge, 79-93.

10. Da Graca Mouga Pocas Santos M (2003) Religious Tourism: Contributions Towards a Clarification of Concepts. In: McGettigan F and Edwards C (eds.) *Religious tourism and pilgrimage*, tourism Board of Leiria, 27-42.

11. Davidson JW, Hecht *A and* Whitney A H *(1990)* The Pilgrimage to Graceland. In: Rinschede F and Bhardwaj S M (eds.) *Pilgrimage in the United States,* Berlin: Dietrich Reimer Verlag.*229-252.*

12. *Deutsche Welle Persian (2012) Just one fifth of the Iranian Imāmzādeh are registered [tanha yek panjom emamzadehaie iran shenasnameh darand in Persian],* Available at http://www.dw.de/%D8%AA%D9%86% D9%87%D8%A7-%D[...]a-16452533

13. Digance J (2003) pilgrimage at contested sites: *Annals of Tourism Research*, 30(1): 143-159.

14. Eade J and Sallnow MJ (1991) *Contesting the Sacred: The Anthropology of Christian Pilgrimage*. London and New York: Routledge.

15. Freyer W (2011) *Tourismus: Einführung in die Fremdenverkehrsökonomie*: Oldenbourg Wissenschaftsverlag.

16. Goeldner CR and Ritchie BJR, (2002) *Tourism: Principles, practices, philosophies*, 9th Edition: John Wiley & Sons.

17. Hochgatterer A (1996) Kulturtouristische Angebotsformen. Am Beispiel des oberösterreichischen Busreiseveranstaltermarktes. In: Tourismus Panorama.

18. Jahrgang. Nr. 3/96, Wien: Forschungskreis für praxisorientierte Tourismus- und Freizeitwissenschaft.

19. Hosta N and Limon B (2009) Changing features of the concept of pilgrimage: the example of the Mevlana's museum in Konya. In: Tore A, Dahla B (eds.) *Pilgrimages Today: based on papers read at the symposium on pilgrimages today held at Åbo*, Finland, 19-21.

20. Jackwoski A and Smith VL (1992) 'Polish pilgrimage- tourists', *Annals of Tourism Research*, 19: 92-106.

21. Jafari J and Ritchie JRB (1981) Toward a framework for tourism education – problems and prospects: *Annals of Tourism Research* 8 (1), 13-34.

22. Morinis A (1992) Introduction: the territory of the anthropology of pilgrimage. In: Morinis A (ed.) *Sacred journeys: the Anthropology of Pilgrimage*, Westport, CT: Greenwood Press, 1-28.

23. Murphy PE (1985) *Tourism: A Community Approach*, London: Routledge.

24. Olsen DH and Timothy DJ (2006) Tourism and religious journeys. In: Olsen D and Timothy DJ (eds.) *Tourism, Religion & Spiritual Journeys*, London and New York: Routledge, 1-21.

25. Olsen DH and Timothy DJ (2006) *Tourism, Religion & Spiritual Journeys*: Routledge Cahpman & Hall.

26. Rinschede G (1988) The Pilgrimage Center of Fatima/ Portugal. In: Rinschede G and Bhardwaj MH (eds.), *Pilgrimage in World Religions: Presented to Prof. Dr. Angelika Siever on the Occasion of Her 75th Birthday*: Dietrich Reimer, 65-98.

27. Rinschede G (1999) *Religionsgeographie.* [*Religion Geography*] Braunschweig: Westermann.

28. Rinschede G (2000) Religious Centre. In: Jafari J (ed.):*Encyclopedia of Tourism*, Berlin: Routledge, 37.

29. Rinschede G and Bhardwaj SM (1988) *Pilgrimage in World Religions: Presented to Prof. Dr. Angelika Siever on the Occasion of Her 75th Birthday*. Berlin: Dietrich Reimer.

30. Rinschede G and Bhardwaj SM (1990) *Pilgrimage in the United States*, Berlin: Dietrich Reimer Verlag.

31. Schmude J and Namberger P (2010) *Tourismusgeographie* [*Tourism Geography*]. WBG.

32. Shackley M (2006) Empty bottles at sacred sites, Religious retailing at Ireland's National Shrine. In: Timothy D, Olsen, D (eds.) *Tourism, Religious and Spiritual Journeys,* USA and Canada: Routledge, 94-104.

33. Sharpley R (2012) Tourism, Religion, and Spirituality. In: Tazim J and Robinson M (eds.) *The SAGE Handbook of Tourism Studies:* SAGE, 237-253.

34. Sharpley R and Sundaram P (2005) Tourism: a sacred journey? The case of Ashram Tourism, India. *International Journal of Tourism Research*, 7: 161- 171.

35. Smith VL (2001) Stone age to Star Trek. In: Smith V and Brent M (eds.) *Host and Guests Revisited: Tourism Issues of the 21st Century.* Cognizant Communication Corp, New York: Putnam Valley, 15-27.

36. Sommer A (2012) Spiritueller Tourismus als eigenständige Tourismusart? Versuch einer Definition und Einordnung [spiritual tourism as a distinct type of tourism? attempt at a definition and classification]. In: Hopfinger H, Pechlaner H, Schön S and Antz C (eds.) *Kulturfaktor Spritualität und Tourismus*, Göttingen: ESV, 21-46.

37. Stausberg M (2011) *Religion im modernen Tourismus*. Berlin: Verlag der Weltreligionen.

38. Stronach D and Royce W R (1981) 'Standing Stones in the Atrek Region: The Halat Nabi Cemetery', *British Institute of Persian Studies*, 19:147-150.

39. Tribe J (2009) *Philosophical issues in Tourism*: Channel view publications.

40. Williams S (2009) *Tourism Geography: a new synthesis*. New York: Routledge.

41. WTO (1995) *Technical Manual: Collection of Tourism Expenditure Statistics.* Spain: world tourism Organization.

42. WTO (2011) *Religious Tourism in Asia and Pacific*. Spain: world tourism Organization.

43. Zamani-Farahani H and Hendersom JC (2011) Iran, Shia Pilgrimage & Tourism. In: *Religious Tourism in Asia and the Pacific,* World Tourism Organization (UNWTO), 163-175.

Chapter 8

Spatial Planning of Rural Tourism with MAPPAC Technique. Case Study Khur and Biabanak County, (Iran)

Faraji Sabokbar,H.A.*Salmani,M.†Ghassabi,MJ.‡Khalvati

University of Tehran (Iran)

*Ashournejad,Q. ** Islamic Azad University,Central Tehran Branch (Iran)*

ABSTRACT

Reviewing the concepts of space and tourism industry, tourism is in an old, deep, unbreakable bound with spatial and physical dimensions. In this way, the lack of systematic and scientific ranking process in spatial locating of rural tourism spots and also improper distribution of infrastructures are the critical deficiencies in this field. The research intends to introduce the hidden potentials and unique capabilities of Khur and Biabanak County, Iran. And prioritize their tourism spots. So tourism planners would be able to recognize proper space distribution. First, the weights of each criterion were calculated by a pairwise comparison questionnaire of AHP method, and MAPPAC technique was used for ranking. AHP was done in Expert Choice software and MAPPAC in MS Excel. Results showed that villages such as Bayaze, Jandagh, Mehrejan, Garmeh, and Iraj which are also older have a higher rank.

Keywords: Tourism, MAPPAC Technique, Khur and Biabanak County, AHP, Spatial Planning

INTRODUCTION

Considering the numerous problems that developing countries faced in achieving top technologies and development these days, tourism could be an appropriate alternative to succeed (Bojanic, 2011, 310). The purpose of tourism development in national and regional level is to equilibrate regional inequality while the main purpose in local level is to balance structural changes and improve local economy (Kauppila et al, 2009, 226). Creating and developing new spaces through tourism development will arrange groundwork for employment and remuneration. The concept of space and spatial analysis as a framework for tourism locating are so important because an appropriate tourism space would influence the region and boosts its economy and its cultural development (Shamaei et al, 2011, 25). Tourism development in any spaces, cities or villages is depending on attractions. Bearing in mind financial provision, constructional evaluation and maintaining policies for tourism spots are very important before any kinds of planning (Salehi fard, 2011, 161). Tourism activities have a great deal in regional planning in recent decades. One of the main problems in spatial development, especially tourism development is the weaknesses of mutual based hierarchy in tourism regions. Therefore, leveling based on tourism infrastructures and available services must be prepared.

Khur and Biabanak County is located in a boundary line of country's central desert and it has great historical and cultural backgrounds. These features bring the region into an excellent term of historical-archeological, social-cultural, and especially desert tourism. The geographical location and historical-cultural background make the tourism activity in this region considerable, and attracted many local and foreign tourists in recent years. But, which villages should visit as there is always the matter of time limitation. In other words, which villages have a better situation in terms of tourism infrastructures and facilities? Therefore, it is necessary to apply systematic approach, using multi- criteria evaluation model to combine all the effective factors and make known tourism priority of the region. The objectives could be defined as follow:

To identify and evaluate tourism development potential specially in the field of desert tourism and historical-cultural tourism,

- To rank rural settlements, according to tourism attractions and infrastructure,

- To prioritize rural settlements in order to optimize servicing in tourism activities.

STATEMENT OF PROBLEM

It is acknowledged that tourism industry in Iran has not met the expectation of making money and improving economic condition which the ignorance of rural tourism is one of the reasons (Ghafari et al, 2011, 98). Iran is among the 10 top countries in natural sites. However, according to the available statistics, this opportunity has not been considered appropriately (Abolhasani nejad, 2003, 2). Tourist attracting procedure needs a deep evaluation and understanding about the space. And lack of ranking and leveling of infrastructures and services are among the obvious and basic shortages which are tangible in so many plans and tourism development projects in our country. Environment as a main pillar of tourism industry has an important role to attract tourist in a space outlook. In order to develop rural societies, problems and structural-functional deficiencies need to be introduced (Ghafari et al, 2011 98). So it could be mentioned that it is essential to determine the preferable tourism spots and ranking them. It causes better servicing and acquires economic and social justice (Shamaei et al, 2011, 24).

Khur and Biabanak County is among the regions which has a variety of plants, animals, geomorphological, historical and cultural attractions especially for desert tourists, but it has not been engaged for sustainable development so far. To take an efficient step toward appropriate planning and management, spatial model must be provided. And providing strategies, solutions and also identifying the groundwork of rural tourism development could revolutionize tourism development in this region.

LITERATURE REVIEW

Few articles have been published up to 1990s in the field of leveling tourism regions. But in a few decades, researchers tried tourism leveling in a regional, national, international scale. In national scale tourism researchers have done the researches in a regional divisions and offer solutions with identifies main tourism spots. Tourism master plan provided by Plan and Budget Organization could be mentioned as one of them (Shamaei et al, 2011, 24). A list of studies that have been done so far in this area is listed in the following table (table 1).

Table 1a. Studies done in leveling tourism regions (2011/2010)

author	year	title	Topic and findings
NahidSajadyan, MahyarSajadyan	2011	Capabilities of rural ecotourism in Mazandaran (zoning, management, maintenance and development)	Analyzing and categorizing the cities of Mazandaran province in terms of rural ecotourism potential
Sayed Amin Ghafari, Mahmud Moradi, DavoudNikbakht	2011	leveling and spatial planning of rural tourism in the central part of the Boyer Ahmad city	Analyzing the natural, historical and cultural potential in rural regions, as well as the way of distribution of spatial tourism services in rural regions and areas
Ali Shamaei, JafarMousavand	2011	Leveling the cities of Isfahan province in terms of tourism infrastructure by using TOPSIS and AHP model	leveling tourism infrastructure based on some of the factors in urban tourism
Eftekhari, Pour Taher	2011	Prioritization of tourism potential in the rural areas of city Nir	Planning of tourism activities in the study area
Zarabi, Mohammadi, Baba khanzadeh	2011	The analysis of local tourism attractions and facilities in Owramanat	Review and identify tourism attractions of the study area and evaluate tourism facilities
Farzin, Nad Ali Pour	2011	Competition advantage factors of tourism destinations (Chabahr region)	Prioritization of the competition advantage factors of tourism destinations in Chabahr region
AmadKhatunAbadi Mehdi RastGhalam	2011	Evaluation of four pillars in rural tourism using SWOT, tourist villages of ChaharMahal and Bakhtiari province	The study of rural tourism status in tourism villages of Chahar Mahal and Bakhtiari Province
MehrdadKarami, Mehdi Modiri	2010	Identifying tourism potential regions, a step towards achieving regional development	The study of feasibility of regional tourism development by studying the potential and infrastructure of a region

Source: the authors

THEORETICAL BACKGROUND

Rural tourism development would be worthy for those kinds of villages which have enough tourism potentials, markets, human resources and enough investment to support tourism section and also the ability to attract public and private sectors.

The situation must response to the following questions before planning procedure:

- Are there enough resources for developing tourism?
- Are there enough markets in a place which could attract tourists to?

- Are there enough efficient human resources to service tourists?
- Do we need tourism to achieve economic development?

Are we able to compete with neighbor countries through current tourism projects? (Ghaderi, 2004, 5).

Table1b. Studies done in leveling tourism regions (2009/1998)

author	year	title	Topic and findings
MasoudTaghvaei, SayedRaminTaghvaei	2009	Spatial planning in tourism development (Chahar Mahal and Bakhtiari, HoorBazoft)	Select a top regional tourism spots and line up functional interactions between them
Ali ZangAbadi, FarahnazAbolhasani	2008	leveling spatial analysis and planning of residential centers by tourism development indicator	A survey to determine the status of development of residential centers to determine the status of units and to provide a qualitative and quantitative to improve centers
Sarah I. Lebermann& Peter Mason,	2002	Planning for leisure and tourism at the local level: Applied Research in New Zealand	Criticize of local planning process and provide a practical model for recreational and tourism planning in New Zealand
Victor B. Teye	1999	Tourism planning and tourism planning challenges in Ghana	Criticize and review tourism development planning in Ghana and choosing goal achievement factors
Wen-Ching Hong	1998	Rural tourism: A case study of regional planning in Taiwan	Description and classification of rural tourism in Natva province which is located in center of Taiwan

Source: The authors

Tourism destinations are very important in tourism marketing procedure. They are major assets in tourism industry. In a condition of unplanned human activities, some hazards and accidents may also occur. Yet, like many other forms of planning, recreation and tourism planning are not simple and clear processes, and some problems can be revealed especially at local levels (Leberman & Mason, 2002, 4). Planning as a scientific tool will help tourism activists to find the best development strategy and to get it in the path of the other section's economic development (Zangi Abadi et al, 2008, 28). Gunn believes that the term "tourism destination planning" refers to a kind of planning for a place where have the essential facilities, attractions, infrastructure and human resources able to attract tourists (Gunn, 1994, 87). The purpose of tourism planning is to identify and supply the needs of tourists (Sajadian et al, 2011, 61). Rural tourism planning and management is a complex matter, because resources are very vulnerable. Natural and man-made attractions, amenities, facilities, marketing, information services, and transportation are all different in rural tourism products. In addition, rural region is a multi-purpose source and tourism is just one of the demands. However, there are two important features in rural tourism: first, the diversity and dispersion of supply and demand and second is the uncertainty feature of demand. These two features will complicate the process of planning and management. Because of the following reasons, take a systematic and integrated approach in rural tourism management and planning appears to be necessary:

The reduction of agricultural activities as a dominant land use and shifting in demands necessitate the policies to allocate resources effectively in other ways such as tourism. Today it is accepted that tourism could be a stimulus for economic and social revitalization for rural regions. Due to the small scale and scattered nature, long-term planning is required to develop rural communities.

Rural resources are fragile so they must meet tourists according to their natural and socio-cultural capacities. Characteristics which can attract tourists to the region are not recognized and it is only achieved from continuous management and planning.

Tourists have high expectation of qualities and diversities of products. And also tourism industry has a long term responsibility toward environment. So, mutual responsibility could only be provided by efficient planning and management. Integrated and comprehensive planning to eliminate deficiencies of existing tourism facilities and services for visitors and locals are necessary (Rezvani, 2008, 159-160). For better planning and marketing of tourism destinations, the concept of tourism destinations should be clear first. In general we can say that the definition of tourism destination contains some elements which could provide a condition to attract and satisfy tourists from the combination and interaction with each other. Each region has some specific attractions and some have natural attractions that cause the extemporaneous development of tourism. Systematic analysis, implementation and updating tourism packages are necessary. And the evaluation of criteria is very important. It includes innate features of a place, organizational and institutional features like space, current capacity measured by tourist's perceptual studies and data analysis. A tourism region leveling is a criterion for determining the center of region and also for determining necessary infrastructures. Nowadays, despite the advancement of computer and statistical techniques in geographical studies, using different is the most common leveling criterion. In this regard, determining and establishing

a hierarchy from tourism regions is necessary and could be an effective framework for dedicating appropriate services and proper performance. In Iran, after finishing the imposed war by Iraq and beginning the first and second development plans, urban tourism in the theoretical and empirical fields was considered, so that in recent years a lot of academic research has been devoted to the study of urban tourism (Shamaei et al, 2011, 24). As previously mentioned, few studies have been undertaken in this area up to now for example there is a research under the title of leveling and spatial planning of rural tourism in the central part of Boyer Ahmad city which is done by Ghafari et al in 2011. The research is done in order to level and determine rural spaces for tourism development, using documentary method, Guttmann model and statistical techniques. In another research under a title of leveling Isfahan province in terms of tourism infrastructure, Shamaei et al (2011) do the leveling process for a region according to some tourism attraction factors by using AHP and TOPSIS models. As there are a lot and various effective criteria in this field, using multi-criteria decision model is taken in order to enhance the accuracy of the calculation and criterions assessment.
Methodology

This study is an empirical research and descriptive-analytical method is used to survey. Considering the purpose of research, effective criteria and indicators to select tourism destination were identified first. After an evaluation from relevant experts and scholars in this field, the indicators reanalyzed and their validity was verified. Following this step, necessary data was collected from accomplished field studies according to the criteria, local authorities and experts' ideas and the existing information sources (census of villages and census of population and housing in the years from 1996 to 2006). After completing the pairwise comparison questionnaire by experts, the weight of each criterion is identified. And (MAPPAC) technique was used for ranking. In this research, Expert Choice software was used to implement the Analytical Hierarchy Process (AHP) and Excel software was used for ranking techniques using MAPPAC.

MAPPAC TECHNIQUE

MAPPAC technique was introduced by Matarazo initially. Then it was used in various fields. MAPPAC technique is a method for ranking multi-purpose

MODM. In this method, all the criteria, in a binary form are generated on the relations between P (preference) and I (indifference) and the rankings that are already done.

The major feature of MAPPAC technique is the presentation of multi-criteria ranking index based on measurements near the ideal solution for classification options (Matarazzo, 1986; Martel and Matarazzo, 2005; Erdal Dincer, 2011: 7). This technique has three basic assumptions (Matarazzo, 1990; Erdal Dincer, 2011: 9):

- Value of v_{ij} for each a_j parameter is determined based on criterion performance .
- The quantity of each v_{ij} can be determined by each K_i criterion for each item of a_j
- .The quantity of v_{ij} from each v_{ij} could be initialized in the range of [0, 1].

The value of v_{ij} is determined for each K_i which indicates the function of v_{ij} based on K_i . A numerical weight ω_i for each K_i which indicates the importance of is determined by $\sum_{i=1}^{n} \omega_i = 1$ (equation 1).

- A function of 7 is created for each K_i and in order to determine the value of υ (v_{ij})
- for each v_{ij} , the following is used: υ0£vij)£1. Based index priority is calculated between each pairs of ω_e and ω_f based on any pairs of K_g and K_h criterion with (Matarazzo, 1991; Erdal Dincer, 2011: 10).

$$_{gh}(\omega_e, \omega_f) = 1 \text{ if } \upsilon(v_{ge}) > \upsilon(v_{gf}) \grave{U} \upsilon(v_{he}) > \upsilon(v_{hf}) \pi$$

$$_{gh}(\omega_e, \omega_f) = 0 \text{ if } \upsilon(v_{ge}) < \upsilon(v_{gf}) \grave{U} \upsilon(v_{he}) < \upsilon(v_{hf}) \pi$$

$$_{gh}(\omega_e, \omega_f) = \tfrac{1}{2} \text{ if } \upsilon(v_{ge}) = \upsilon(v_{gf}) \grave{U} \upsilon(v_{he}) = \upsilon(v_{hf}) \pi$$

$$\pi_{gh}(\alpha_e, \alpha_f) = \frac{\omega_g(\upsilon(v_{ge}) - \upsilon(v_{gf}))}{\omega_g(\upsilon(v_{ge}) - \upsilon(v_{gf})) + \omega_h(\upsilon(v_{hf}) - \upsilon(v_{he}))} \text{if}$$

$$(\upsilon(v_{ge}) > \upsilon(v_{gf}) \wedge \upsilon(v_{he}) \leq \upsilon(v_{hf})) \vee (\upsilon(v_{ge}) = \upsilon(v_{gf}) \wedge \upsilon(v_{he}) < \upsilon(v_{hf}))$$

$$\pi_{gh}(\alpha_e, \alpha_f) = \frac{\omega_h(\upsilon(v_{he}) - \upsilon(v_{hf}))}{\omega_g(\upsilon(v_{gf}) - \upsilon(v_{ge})) + \omega_h(\upsilon(v_{he}) - \upsilon(v_{hf}))} \quad \text{If}$$

$$(\upsilon(v_{ge}) \leq \upsilon(v_{gf}) \wedge \upsilon(v_{he}) > \upsilon(v_{hf})) \vee (\upsilon(v_{ge}) < \upsilon(v_{gf}) \wedge \upsilon(v_{he}) \geq \upsilon(v_{hf}))$$

π_{ef} is determined by the following, (equation 3):

$$\pi_{ef} = \sum_{i<j} \pi_{ij}(\alpha_e, \alpha_f) \frac{\omega_i + \omega_j}{m-1}$$

And a general value π_e is determined by following equation for the item α_e. (Equation 4):

$$\pi_e = \sum_{\alpha_f \in A \setminus \alpha_e} \pi_{ef}$$

Then, α_e is selected with a biggest π_e and it is also selected as an optimal item. π_e is recalculated with a biggest allocated value of π_e which is selected as a best second alternative not including the optimal option of A and the remaining value of α_e.This process is repeated until the ranking of all alternatives then a similar process begins with the selection of the least optimal choice of A, the choice is removed from A and π_e is recalculated again. And the remaining value of α_e with a lower value of π_e is selected as a best second alternative the process continues until the rank of all the alternatives have done. The ascending and descending ranking combine in order to achieve a weak ranking of A.

GEOGRAPHICAL LOCATION OF THE STUDY AREA

Khur and Biabanak County is located in Isfahan province. It has three towns called Biabanak, Mehrjan, Nakhlestan and 25 villages and 201 farms in the southeast hinterland of Dasht-e Kavir (a large desert in Iran). It is bounded in the north by the Semnan Province, by Yazd in south and in east by Khorasan province. It is 400 kilometers far from the capital of province. It has an extent of 11600 square kilometer. The county has the population of 21000. And 8000 of them live in a center of the county (census of villages

2006). Khur and Biabanak County is considered as one of the most attracting tourism areas because of its abundant tourism attractions (table 2). In this research we try to rank rural settlements in terms of tourism leveling process.

Table 2: Khur and Biabanak tourism attractions

Type of attraction	item	cases	
Natural attractions	geotourism	Mountains - salt - marsh sand - sand hill	Playa Khur in East and North East Khur, Jinn sand, pebbles Ashtian, sand dunes in rural Mesr springs spa (Spa Village) and Bazyab mineral spring saline lakes, saline desert and sprin, subterranean Klaghv with therapeutic properties.
	Plant species	Old trees	Ayraj cypresses, Ardyb, Khnj (to see the traditional architecture of the mentioned villages, cypress trees and village houses view to cypresses, ancient olive trees Byazh, Jandagh
		Mastic communities - bitter almond, Qych, Artemisia, Ashnan, tamarisk, reeds, Haloxylon.....	
	Animal species	Hubreh, partridge, rabbit, fox, whole, cheetah, wild cats, lizards (18 species)	
Historical attractions	castles	Byazh Castle, Ayraj Castle, Garmeh castles, Jandagh castle	
	caravansary	Anarak, Rabat behind Badam, Ali Abad, Rabat Khan, Moshajari (the old way)	
	houses	Ebrahimi in Anarak, Yaghma in Jandaq, Mazyar in garmeh, and Hussain Khan and Karbasi in Byazeh, and Haji Hussein Agha, Pazarv in mhrjanHashemZaye in Byazh Divan Khane in Ardyb ...	
	Water reservoirs and subterranean	Byazh subterranean, Khur, Farm Nobahar,	
Social and cultural attractions	handicrafts	Palm Related industries (cradle, Aquarius, Libra), or camel ...	
	Ethics and traditions	Palm battalion (Byazh and Khur) New Year celebrations, a woman religious mourning (Byazh), ritual ceremonies	
Others	Observations of stars in the last and the first night of each month, seasonal springs, deserts and some types of queries (paste, salt, black), sacrifice and beheading of a palm tree, a camel and agricultural practices ...		

THE CHOICE OF TOURISM DESTINATIONS, EFFECTIVE INDICATORS AND CRITERIA

As a consequence of progressive use of computer and statistical techniques in geographical studies, using different indicators in the various fields of tourism areas is the most common leveling criterion these days. In this regard, it is necessary to determine a hierarchy of the regions which would be able to provide an effective framework for appropriate services, distribution and function (Shamaei et al, 2011, 24). Tourism destinations are always ranked according to health and motivational factors. And the highest factors among the ranking factors are safety, tranquility and non- infected environment. Since these criteria exist more or less in rural areas, they will be considered as tourism attraction factors (Rezvani, 2008, 66). Edwards divided and distributed effective variables in tourism attractions in a rural area into two separated factors: public and socio-cultural (Edwards, J. et al, 2000, 28). Salehi Fard (2011) and Ghaderi (2004) described tourism attractions and resources in terms of man-made attractions, cultural attractions and natural attractions. Lamsdn considers four elements for a tourism destination and its classification (Table 3).

Table 3: Lamsdn classification elements for tourism destination

features	description	cases
Primary attractions	major attractions which attract tourists and makes the difference from one destination to another	Acropolis in Athens, the pyramids of Egypt, the TajMahal in India
environment framework: visual identity	Physical design of a destination including: beach resort, the historical sites and business places. The main elements including: the infrastructures, road and rail networks, open space and community facilities	The beach of Boston Wharf area of London, Venetian channels, Roman Quarter in Paris
Supportive services	facilities such as accommodation and lodging, communications, and transportation, beverage and food, entertainment and amenities	Essential for all cases
Socio-cultural dimensions	Cultural characteristics, the bridges between the past and the present, comfort and tranquility, the warmth and solidarity between tourists and the host community	Mass traffic in Delhi India, music in Havana Cuba, intimacy in Greece islands

To locate a tourism destination, following matters should be considered: proximity to the other recreational sites, the attractiveness of the location, climate condition, beautiful environment, having enough land for future development and extra buildings, access to freeways and recreational places, adequate public facilities and transport infrastructure, such as water, electricity and telephone, immune environment of the region, residents hospitability, residents good attitudes toward tourism, adequate and professional human resources (Heydari, 1387, 185 ; Ghaderi, 1383, 57).

Since rural tourism is a complicated and multi-lateral activity which involves different types of farm-based trips, ethnic, educational and ecotourism trips, efficient and harmonized planning and management in rural tourism is a complicated matter (Sajjadian et al, 1390, 61)

Investigating the presented indicators in the survey, the index of the table (4) has been derived. Finally, despite the vast extent of criteria and indicators and the absence of some of indicators in the Census of Population and Housing (1385), data and information were collected by field study methods. (Table 5)

Table 5. Ranking results of Khur and Biabanak County with MAPPACK

village	rank	village	rank	village	rank
Byazh	1	Jafarabad	11	Abadan	18
Jandagh	2	Haftooman	11	Neyshabur	19
Mehrjan	3	GolestanAroosan	11	Aroosan Kure Gaz	20
Garme	4	Ebrahim Abad	12	Jegarag	21
Iraj	5	Farahzad	12	Mohammad Abad Kure Gaz	22
Mesr	6	Hossein Abad	13	Aziz Abad	22
Farkhi	7	Qadrabad	14	Quarry Complex of Marmarit	23
Ardib	8	Nasr Abad	15	Abgarm	23
ChahMalek	9	Amir Abad	16	Hajrag	24
Khanj	10	Bazyab	17	Behin	25

Reviewing the history of residency in the city of Khur and Biabanak and also considering undertaken spatial analysis in this study, it is clear that rural settlements with archaeological origin ranked higher than others like Byazh, Jandagh, Mhrjan, Garme and Ayraj; This is due to the historical and cultural attractions in these settlements; And it would be more attractive to tourists, especially those who are interested in cultural and historical attractions. Villages with these features could be good destinations for tourists interested in historical and cultural attractions. But the point is the capability and potential of region to attract tourists and providing services for them. Because of climatic conditions and a vast expanse of the area and the far distance to other destinations, the shortages of amenities and facilities for tourists are strongly felt. Considering the total conditions, the requirement allocation for infrastructure facilities and services is identified from the selected indicators. The results can be a proper criterion for allocating funds and facilities to the region.

CONCLUSION

To get hold of a great opportunity in the tourism industry opening up before in the near future, Bangladesh's tourism sector must start taking all the preparations from this moment. At the same time it must not also ignore the increasing prospect of the domestic market of this sector. As mentioned earlier, it is not enough that the country possesses a potential for becoming a covetable tourist destination. To turn that possibility into reality, marketing is a pre-condition. Marketing strategy of the tourism sector must have a clear attitude for implementing the strategy. Tourism sector should simultaneously utilize its assets and abilities in the real life scenario of the country to add value. High quality products at a low price, arrangement for entertainment, development of infrastructure, security, accommodation, aggressive promotional campaign etc. are also required. Growth of the tourism industry largely depends on national and global condition. Pricing strategy of the tourism sector is very important for which all categories of tourists including high-income group and low-income group may get the opportunity to tour. Product differentiation technique should be used to attract tourists and fulfill their demands. Tourism industry should be comprised of accumulating all courses of action for which efficiency and effectiveness in the strategic marketing is being required.

REFERENCES

1. Abolhasani Nejad, B., (2003), Capabilities and limitations of ecological tourism in the city of Ramsar and Tonekabon, MA Thesis, Department of Geography, Tehran University.

2. Badri, A., Yari, A., (2009), selecting tourism spots using AHP, case study, Kohkiloyeh and Boyer Ahmad, Geographical Research Quarterly, No. 95.

3. Bojanic David, (2011): Using a tourism importance–performance typologyo investigate environmental sustainability on a global level, Journal of Sustainable Tourism,DOI:10.1080/09669582.2011.584624.

4. Edvards,J. et al,(2000) Tourism brand attributes of the alto minho, Portugal, in tourism and sustainable community development, edited by Derek Hall and Greg Richards, Rutledge Advances in Tourism, Lonlon.

5. Eftekhari, A., Pourtahery, M., Mahdavian, F., (2011), prioritizing rural tourism in the city Nir, Geography and Development, No. 24, 23-38.

6. Erdal Dincer Sait, (2011), the Structural Analysis of Key Indicators of Turkish Manufacturing Industry: ORESTE and MAPPAC Applications, European Journal of Scientific Research, Vol. 60, NO.1, 6-18.

7. Farzin, M. Nad Ali Pour, Z. (2011), Factors affecting competitiveness of tourism destinations in Iran, Chabahar case study, Journal of Tourism Studies, No. 14.

8. Ghadri, Z. (2011), the principles of sustainable development of rural tourism planning, municipality publication.

9. Ghaffari, R. , Moradi, M., Nik Bakht, D. (2011), the leveling and planning of rural tourism space in centralBoyerahmad city, urban and regional studies, third year, No. eleventh.

10. Guun, C. A. (1994) Tuorism Planning: Basic Concepts, Cases, Washington. D. C: Taylor and Francis.

11. Heidari, R. (2008), Principles of Tourism Planning, Samt publication.

12. Karami, M., Modiri, M. (2010), identify potential areas of tourism, a step towards achieving regional development (Kalpourgan case study), Journal of Geographical Sciences and Applied Research, Volume 14, Number 17.

13. Khatoon Abadi, A.,Rast Ghalam, M., (2011), measuring the four pillars of rural tourism using SWOTtechniques, tourist villages of Chahar Mahal and Bakhtiari, Journal of Development Economics, Volume 25, Number 3, 330-338.

14. Leberman S. I., Mason P. (2002): Planning for recreation and tourism at the local level: Applied research in the Manawatu region of New Zealand, Tourism Geographies, 4:1, 3-21.

15. Martel, J. M., and Matarazzo, B. (2005), "Multiple Criteria Decision Analysis: State of The ArtSurveys", International

Series in Operations Research and Management Science, Volume 78,III, 197-259.

16. Matarazzo, B. (1990), "A Pair Wise Criterion Comparison Approach: The MAPPAC and PRAGMA Methods", Readings in Multiple Criteria Decision Aid (ed. By Bana e Costa, C.), pp. 253-273, Springer, Berlin.

17. Matarazzo, B. (1991), "MAPPAC as A Compromise Between Outranking Methods and MAUT", European Journal of Operational Research, 54, 48-65.

18. Matarazzo, B. (1986), "Multicriterion Analysis of Preferences by Means of Pairwise Actions and Criterion Comparisons (MAPPAC)", Applied Mathematics and Computation, 18 (2), 119-141.

19. Pekka Kauppila, Jarkko Saarinen & Riikka Leinonen (2009): Sustainable Tourism Planning and Regional Development in Peripheries: A Nordic View, Scandinavian Journal of Hospitality and Tourism, 9:4, 424-435

20. Rezvani, M. (2008), the development of rural tourism with sustainable tourism approach, Tehran University Press.

21. Sajadian, N., Sajadian, M. (2011), ecotourism potential in Mazandaran (zoning, management, maintenance and development), urban management, No. 27.

22. Salehi Fard, M., (2011), rural tourism, principles of planning and design of structural projects, Marandiz publications.

23. Shamaei, Ali, Musavand, J. (2011), leveling the city of Isfahan in terms of tourism infrastructure by using TOPSIS and AHP, urban and regional studies, third year, No. eleventh.

24. Sinha, P. C., (2011), World Tourism, Sustainable Ecotourism, translated by Mohammad Quli Pour, Katibe Publications.

25. Statistical Center of Iran (2006), Census of Population and Housing, 1385. Statistical Center of Iran (2006), Census of Villages, 1385.

26. Taghvaye, M., Ghaffari, R., (2010), spatial planning in tourism development (case study, Chahar Mahal and Bakhtiari province, Bazoft), Geographical Research Quarterly, No. 96.

27. Victor B. Teye (1999): Commentary: Tourism plans and planningchallenges in Ghana, Tourism Geographies, 1:3, 283-292

28. Wen-Ching HongDivision Chief(1998), Rural Tourism: A Case Study of Regional Planning in Taiwan Agricultural Bureau,Nantou County Government 660, Chung Hsing Rd., Nantou City, Taiwan, 1988-10-01

29. Zangi Abadi, A., Abolhassani, F. (2008), Spatial Analysis of leveling and resorts planning with Tourism Development Index (TDI), University Jahad Daneshgahi.

30. Zarrabi, A., Mohammadi, J., Babakhanzadh, E., (2011), Analysis of local tourist attractions and facilities in Oraman region, Journal of Geography and Environmental Planning, Year 22, No. 3, 35-52.

Chapter 9

Shariah Compliance Hospitality Building Design: A Malay Muslim oriented architecture

Noor Hanit[a] Abdul Majid[a]*, Zuraini Denan[a], Fauziah Hanum Abdullah[a], Mohd Syukri[a] Mohd Noor[a]

[a]Department of Architecture,Kulliyah of Architecture and Environmental Design International Islamic University Malaysia,IIUM.

ABSTRACT

As a Malay Muslim architecture, the traditional houses suggest characteristics of Syariah compliant (SC) design. In addition, assimilation of traditional Malay architecture to the hospitality buildings is seen as an approach to promote cultural tourism. Hence, this research aims to investigate the influences of traditional houses on the design of hospitality buildings. The methodologies adopted are case studies and literature reviews. The study analyses the SC characteristics of the Malay houses, discusses the SC hotel requirements and review selected examples of SC hotels as case studies. The findings indicated that there are profound vernacular features which compliments Syariah.

Keywords: Syariah compliant; hospitality building; Malay traditional house

INTRODUCTION

In Malaysia, a Malay person has synonymously been accepted as a Muslim. This status is defined in Article 160 of the Constitution of Malaysia. The article describes a "Malay" as a person who professes the religion of Islam, habitually speaks the Malay language and conforms to Malay customs. Whilst, the traditional architecture of the Malay people is manifested in a traditional Malay house that hasbeen coined as a reflection of the Malay culture and way of life. Due to the faith of the Malays, the traditional Malay house also reflects and relates to the norms and governance of Islam. As a MalayMuslim architecture, the traditional houses suggest characteristics of Syariah compliance (SC) design. For example, the segregation between genders, female and male domains and visual privacy considerations in the design of the traditional houses. Hence, it is anticipated that the architecture of the houses support and uphold the requirements of Syariah. Surah Ghafir (verse 82) suggested that there are evidences to be discovered if one traces and studies a given environment.

"Have you not traveled through the earth and seen what was the end of those before them? They were more in number than them and mightier in strength, and in the traces (they have left behind them) in the land; yet all that they used to earn availed them not" *(Surah Ghafir, 82)*

In line with the hospitality industry, Malaysia offers a lot of tourism opportunities concerning the culture diversity, multiracial society, and ethnic minority. Tourists can find the whole of Asia in one country as the tagline goes; 'Malaysia Truly Asia' (Malaysian Airlines). Most of the time, tourists are inclined to find out the unique characteristics of the countries they visited. In realizing this matter, Sani and Mahasti (2013) explored the importance of 'regional identity' to a hospitality project in Iran and discovered that hospitality development with national identity increased the number of tourists.

The *halal* or SC product is essential besides the demand for cultural identity as mentioned by Zulkifli et al. (2009) and Sani and Mahasti (2013). In addition, a recent study by Kamaruddin et al. (2013) reveals that people are willing to pay for *halal* product and logistics. The willingness to pay for *halal* products indicates the inclination towards preserving halal compliance parameters. Nevertheless, the essence of culture and identity in Malaysia or specifically the Malays is rapidly decreasing due to development and modernization. Developers and builders compete to build contemporary and so-called iconic building and paid less attention to the fundamental nature of Malay or vernacular architecture. Traditional Malay house is part of heritage value that attracts tourists to visit Malaysia and experience the culture. Neglecting this value will cause the country to lose its sense of place.

Syariah is Islamic religious law that governs not only religious rituals, but aspects of day-to-day life in Islam. The day to day life is closely related to the living spaces; i.e. the houses. In addition, the hospitality architecture in Malaysia is seen as the closest approach to introducing foreign visitors to the traditional architecture of Malaysia. Assimilation of traditional Malay architecture to the hospitality buildings is a venue to promote cultural tourism. This attempt will lead to the preservation and projection of the Malay culture in a new development. In addition, the SC hospitality services have increasing demands from tourists. Hence, this paper will discuss the application of the traditional Malay architecture in hospitality buildings as a gesture to comply with the requirements according to Syariah.

OBJECTIVES

Due to the mentioned factors of traditional Malay architecture and SC hospitality building, this research intends to investigate evidence of the Malay traditional architecture characteristics in the design of hospitality buildings in Malaysia. Since cultural tourism is a value that attracts tourists, it is important to include the influential attributes of Malay Muslim architecture (traditional Malay house) into the design of hospitality building. The insertion of the Malay Muslim architecture could provide a new platform for tourists to understand the cultural meaning of Malay lifestyle.

METHODOLOGY

The methodology adopted for this study are literature reviews and case studies. This study focuses on defining the essence of the functional space of traditional Malay houses as well as Islamic values embedded in it. As the zoning and space divisions in the traditional Malay houses in Peninsular Malaysia are similar, this paper discusses a sample that is a common layout typical of the traditional Malay houses in Perak. In addition, this paper also reviews the SC hotel requirements and selected a few examples of SC in Malaysia as case studies.

This paper will provide an overview of the functional spaces in a traditional Malay Muslim house and discuss the spaces in conjunction with Islamic practices. Moreover, this paper will also present the concept of SC hotels to understand the practices required to comply for the status. The analysis will recommend and elucidate the incongruity among the elaborated issues to the application of the Syariah Compliance to Hospitality building design. Nevertheless, the result of the study will contribute towards new circumstances of bringing back the Malay identity into the contemporary hospitality building while increasing the high draw rate of visitors.

The research framework of this study is divided into three (2) categories, which are as follows:

Analysis of Malay Muslim culture and identity (through understanding the essence of traditional Malay house) Analysis ofSyariah Compliant Hospitality Services

ANALYSIS

This section presents the analysis oftraditional Malayhouse and Syariah Compliant Hotel.The traditional Malay house represents a Muslim traditional dwelling; where a Malay is accepted as a Muslim by definition of the Constitution.

Analytical review of Malay Muslim traditional house

According to A. GhaffarianHoseini et al. (2009), Malay vernacular houses, are efficiently adapted not only to the local climate and environment but also to the local culture. The local culture reflects the identity and way of life of Malays. The Malay Muslim occupied traditional Malay houses and practiced the local culture as their norms. Gurupiah and MohdTajuddin (1998) reiterated that the Malay is practicing Islam in their daily life. Apart of that, Abdul Halim and Wan Hashim (1997) defined that Malay customs in traditional Malay house functions and culture is in line with the Syariah and does not contradict with Islamic teachings. Mortada (2003) also declares that as long as traditional Malay house functions and culture does not conflict with Syariah, Islam permits the dwelling to follow the cultural norms and customs.

Despite the different typologies of Malay house according to different regions; the cultural meaning and essence of traditional Malay house design remains similar. The layout and spatial planning of the houses remain analogous for the different typologies. For instance, the division of spaces that are public, semi-public, semi-private and private are the same. It could be seen in the separation of the interior layout for different genders (Abdul Halim, 1983). One of the Islamic principles that focus on space segregation in Malay traditional dwelling is the concept of privacy (Abdul Halim, 1983). Extracting the Islamic values from Malay traditional house can enhance learning for tourists about different cultures of Muslim from different regions.

Thereby, the study intends to realize the Islamic cultural values in traditional Malay houses and suggests appropriate adaptations to Syariah Compliant hotel design in Malaysia. This gesture is important to instill the local lifestyle for tourists to experience during their visit and stay in the resorts, hotels, homestay or others. The following illustration is the outcome derived from the analysis of a sample layout of traditional Malay house (figure 1);

- The layout of a traditional Malay house has several Islamic influences in its design. This Malay house design
- implements concept of Muamalah in Islam or relationship among mankind. It is expressed through the segregation of gender for entrance to the house. Male and female guests used different entrances and spaces. Male guests enter through porch (Serambi) to occupy the verandah (anjung) while female guests use entrance near hallway (selang) and occupy the main space (RumahIbu).
- The segregation of genders also takes place among house occupants. A Malay house will be occupied by a
- family; parents, daughters, and sons. Parents will sleep in the bedroom, daughters, in the main space (RumahIbu) while sons usually sleep in the verandah. This separation is common in Malay traditional houses.
- The level of privacy is an important concern in Malay house. The size of openings, doors, windows and space
- hierarchy play significant roles. The private area that is dedicated for females in the traditional Malay house rises above other spaces. The idea is to control the privacy level of the occupant of the house and female members. In spite of that, the provisions of the louvers on the window's panel are abiding by the privacy safeguarding requirements.

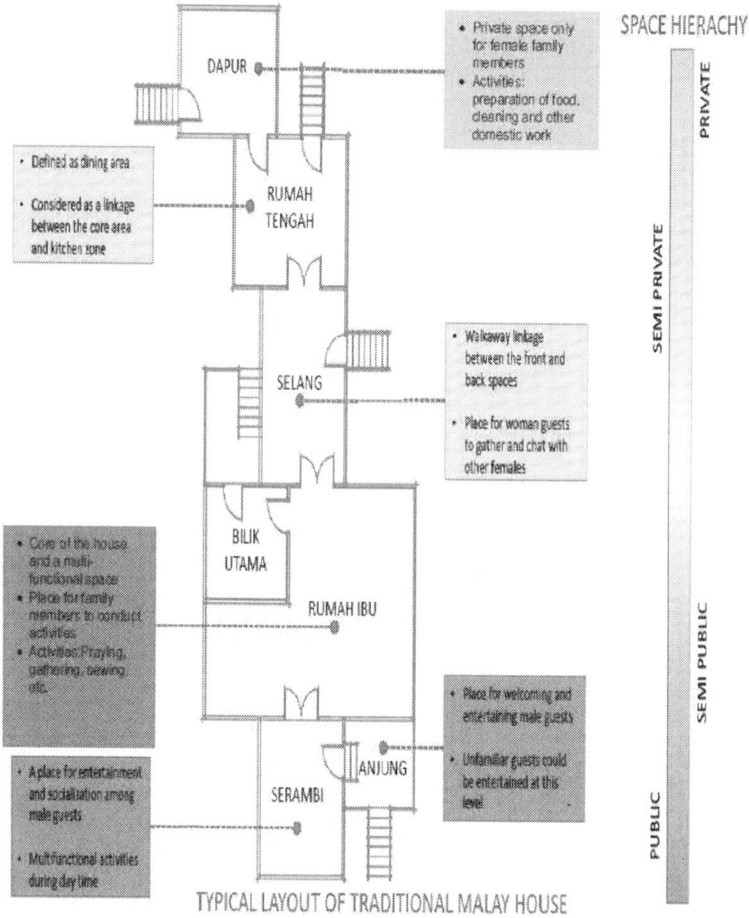

Figure 1. Typical layout of Malay traditional house.

- Toilet planning also takes consideration of the Islamic requirement of orientation and segregation. The bathroom and toilet should not face the Qiblah direction and is detached from the main house.

- The layout of the Malay traditional house is an open planning concept (Abdul Halim, 1983). The interior space

- uses minimal partitions to divide spaces. Therefore, there is a minimal number of bedrooms in the house or none at all. In some cases, temporary partitions are used by hanging cloth. Lim (1987) also declares that the open plan concept is emphasizing the Malay culture tradition shows respect for the elderly, family bonding and the position of female.
- The zones in a Malay house functioned for multiple activities. The spaces are very flexible to the Malay lifestyle

(Lim, 1987). The spaces can be used for multi-functions at different times of the day. The Malays maximizes the space for the activities related to their lifestyle (figure 1).

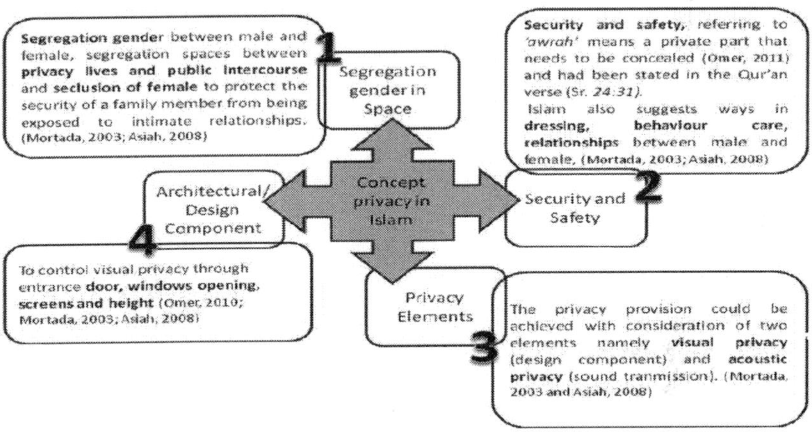

Figure 2. The concept of privacy. Source: Noorul & Anuar (2013)

The analysis of the traditional Malay house outlined some pertinent considerations in designing spaces for Muslims. The considerations include segregation, privacy, space planning, flexible and multiple uses of spaces. In this regard, the privacy value in a Malay Muslim house is seen in the spatial organization, floor plan design and wall opening. This element of privacy is a prime determinant in Malay Muslim traditional house. Visual privacy applications can be used in the form of wall, window, and curtain to block the view from intrusion from a non- mahram eye. Nevertheless, it also can be seen in the spatial arrangement of space that emphasis on segregation gender and separation between public and private space. The understanding of these divisions and treatments toward privacy should be taken into consideration in designing a Muslim dwelling in the future (Noorul and Anuar, 2013). Figure 2 indicates the four main concepts of privacy in Islam as discussed by Noorul and Anuar (2013) that are segregation of gender, security and safety, privacy elements, and architectural design components.

Analysis of Syariah compliant hospitality services

There is a new wave in hospitality services with the booming demand for SC Hotel and Resort. The reason being the significant size of Muslim consumers or travelers and the growing demand for the concept of Islamic hospitality services. The concept of SC hotel and resort have mostly flourished in the Middle-east countries. Based on data from United Nation World Tourism Organization (UNWTO), Malaysia is one of the top destinations for Muslim tourists. Unfortunately, in Malaysia the SC hospitality building design is yet to be developed. Currently, only five (5) hotels claimed to offer SC services; De Palma Group of Hotel, Ansar City Islamic Hotel, Mesra Inn and Al- Taqwa Islamic Hotel and Grand Bluewave Hotel. The SC effort is still in the preliminary stage. For instance, De Palma Hotel Ampang only covers the external aspects i.e. halal food, staff attires, alcohol-free food outlets but lack in the internal aspect (Nur' Hidayah et al., 2012). However, the hotel is recognized as the pioneer SC hotel in the Asia region. Other hotels that claimed as SC hotel mainly cover the operational of halal food that is called Dry Hotel. Apart from these factors, not many hotels assimilate the Islamic essence onto the whole hotel operation, management, and design or in the physical form of the hotel.

SC Hotel can be defined as a hotel where the services offered, and financial transactions are based on Syariah principles. Compliance is not restricted to serving halal food and beverages. An SC hotel should be designed for health, safety, and economics that would benefit all mankind, regardless of race, faith or cultures.

The applications of the SC Hospitality Service should cover operations, maintenance and design of the hotel. Some group of researchers have discussed the attributes of the SC hospitality buildings and services that need to be complied. The model of SC hotel consists of food and beverage, which are non- alcoholic, halal food and beverage with cooking and serving utensils in clean condition. The facilities should take into account the segregation facilities by gender, inclusion of ablution taps in bathroom, comfortable prayer spaces or room, provision of Holy Quran, prayer mats, Qibla direction, Islamic practices booklet, prayer call in the hotel rooms, halal toiletries and prohibit Non-Islamic entertainment. The management should employ Islamic finance and follow zakat principles. The staff attire must cover their *aurah* (body parts as prescribed by Syariah) and practice Islamic attitudes or good *akhlaq* (manners and conduct).

Besides, the hotel must provide clean and comfortable environment. The entrance to the private rooms should not be designed to face one another. The doors and windows should be discreet, offset and not placed adjacent to one another. In addition, the rooms are not recommended to be located on the ground floor. In any cases that the rooms are on the ground floor, the placement of windows is critical. The windows should provide privacy and place at 1.73 meters from the ground and above eye-level. The interior decorations should be void of a human form and encourage the use of Islamic calligraphy art. The bed and toilet should not face the qiblah direction and lastly, there should be a separation between bath and toilet. The entrance to the toilet should have a door that swings to the right so that the user will enter with the left leg. Material choices should allow for acoustical privacy especially between the toilet and the room and rooms. Another important consideration in the building design includes designing with energy efficient and passive design strategies to encourage savings in the operation of the building. Saving energy is in line with Syariah guidelines on savings and preventing wastefulness. The recommendation for SC hotel is illustrated in table 1 based on the practices in traditional Malay houses, Syariah compliant and conventional hotel practices.

Table 1. Recommendations for Syariah compliant hotel design.

NO	MALAY MUSLIM ARCHITECTURE	HOTEL: SHARIAH COMPLIANT HOTEL	HOTEL: CONVENTIONAL HOTEL	RECOMMENDATION
1	Visual Privacy (Opening)			
	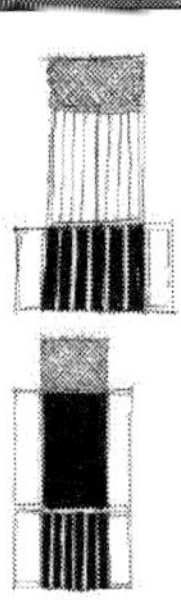	 (De Palma Hotel) 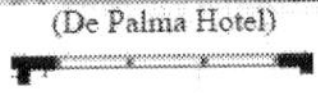	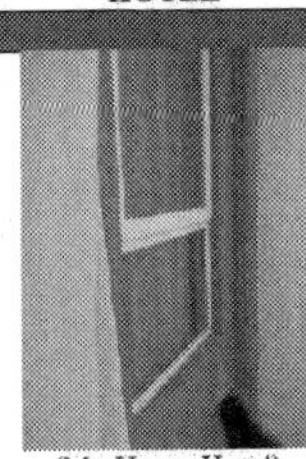 (My Home Hotel) 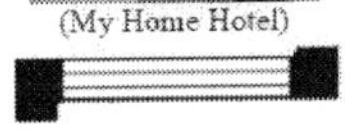	 An example of window that can be applied in many different sizes and proportions.
	Versatile opening design for visual privacy	Glass window with curtain to control privacy	Glass window with curtain to control privacy	Glass window with exterior carving frame to support provision of privacy.
2	Segregation			

	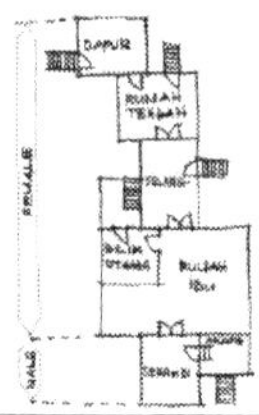	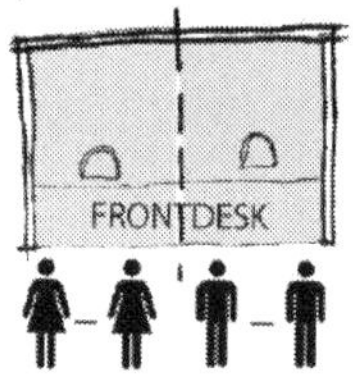		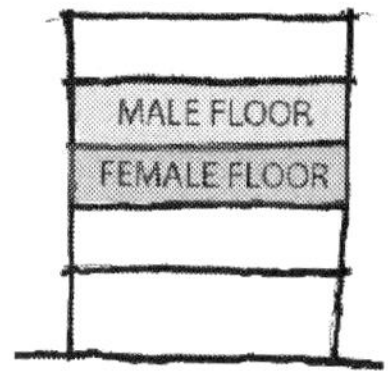
	Different entrance and zoning for different genders.	Gender segregation at front desk service and guestroom floor planning for male and female.	Not available	Gender segregation designated by the floor; i.e. floor designated for male or female guests only or floor for a family to cater for the need of privacy due to the guests' status.
3	**Toilet Planning**			
	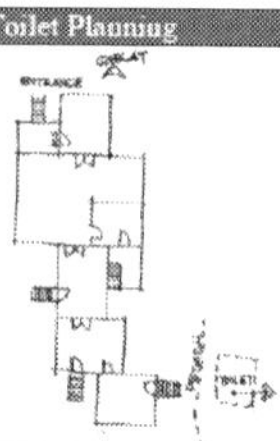			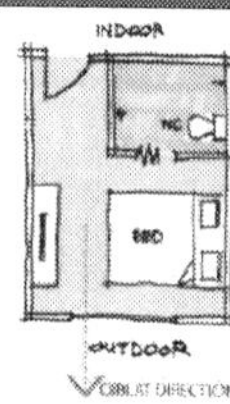
	Toilet orientation is away from qiblat direction. Toilet is also treated as an out-house to maintain the concept of cleanliness.	Toilet orientation is away from qiblat direction.	Toilet orientation is away from qiblat direction.	Toilet (WC) must not face qiblat direction.
4	**Entrance (Implementation of scale or height)**			
	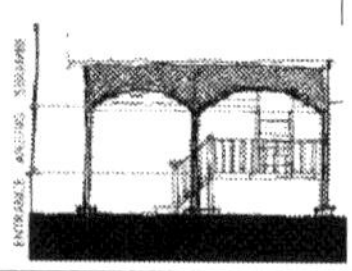	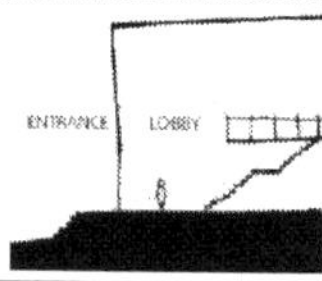		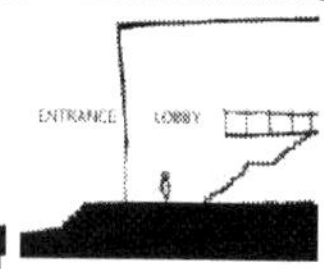
	Anjung and entrance gives sense of welcoming with the hierarchy of steps and space.	Grand, open and welcoming sense for guests.	Standard lobby space.	The lobby area can assimilate the sense of welcoming that accentuates the Malay hospitality.
5	**Multifunctional spaces**			
	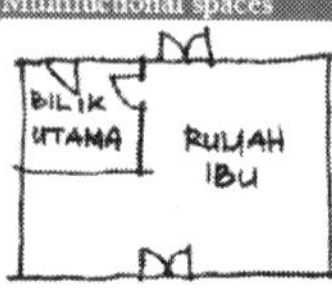	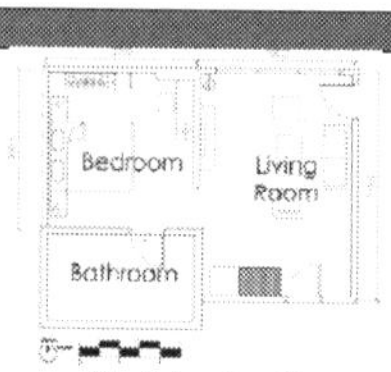 (De Palma hotel)	 (My Home hotel)	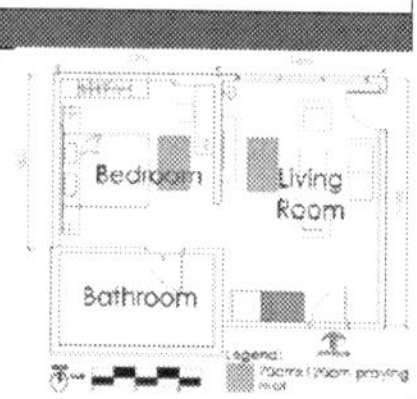
	RumahIbu in Malay house is multifunctional	Insertion of Living Room as a multifunctional area.	Not available (budget room)	Living space in the hotel (guestroom) may function as

6	**Zoning**			
	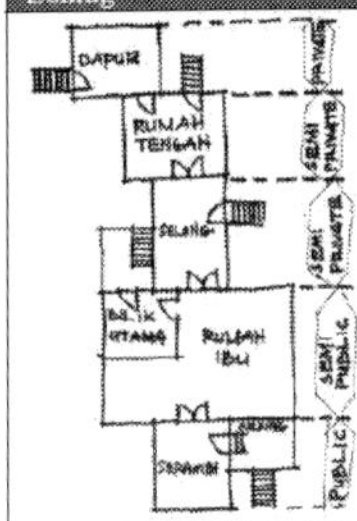	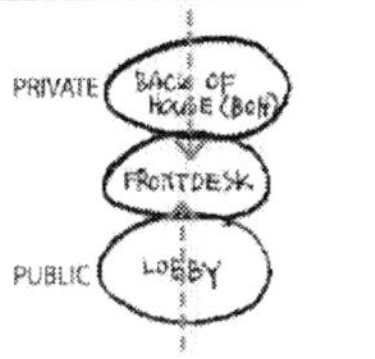	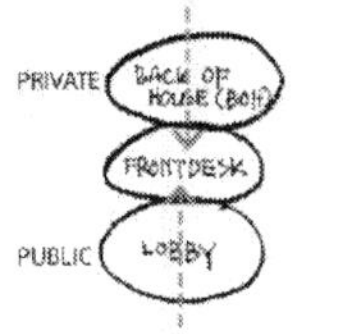	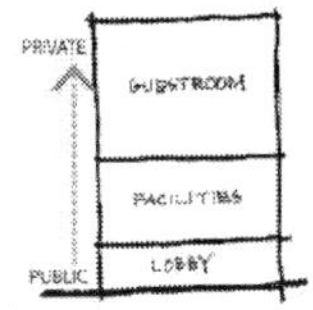
	The traditional Malay house is divided into zones; public, semi-public, semi-private and private	The zoning for the administration in SC hotels.	The zoning for the administration in conventional hotels.	The zoning to be applied to the administration spaces and building design.

FINDINGS AND RECOMMENDATIONS

Resorts, hotels, and homestays are part of the hospitality industry providing accommodation, meals, and other services for travelers and tourists. The analyzes conducted suggest that the accommodation should provide privacy for the users, and proper considerations should be given to the spatial planning, layout, interior and exterior details. For instance, the spatial planning in the hospitality building design should divide the spaces into public, semi-public, semi-private and private zones. The public space in traditional Malay house received and entertained non-mahram guests namely neighbors, relative, and friends. In the hotel context, public space should be shared for all guests, for instance, the lobby area. On the other hand, the guestroom is a private area that is specially created for the guests needs and requirements. Some guestroom may have a living area that can be zoned as semi-public or semi-private area.

The analysis on the Malay Muslim traditional houses and SC hotel have outlined the important considerations to adhere to considering and provision of SC hotel that will reflect the image desired for Malaysia. Thus, in assimilating the conditions prescribed by Syariah law and Malay Muslim architecture the following recommendations are made to explain and elaborate the general Islamic principles occupied in designing a hospitality building with additional Malay essence.

The following are some of recommendation to the designs of the hotel discussed according to the spaces (table

1);

- Lobby

a. Lobby design plays a significant role in selling the hotel hallmarks. It must give a sense of welcoming to the guests. Anjung and entrance area in Malay house functions as what a lobby should be. The layout should be open and flexible to accommodate and entertain the guests. The value of respect should be expressed in this space and can be achieved through the implementation of scale.

- Guestrooms

The guestroom is the most crucial space in the hospitality building as it is the largest income generator. Every hotel has their image, theme and hallmarks on designing the guestrooms. Guestrooms segregation by gender may be designated by the floor; i.e. floor designated for male or female guests only, or floor for the family to cater for the need of privacy due to the guests' status. This segregation of gender is practiced in the provision of different areas in the Malay house. For instance, the serambi and anjung is a male domain while the RumahIbuand dapurare the female domain.

a. Apart of that, *RumahIbu* in traditional Malay house is a multifunctional space that most activities of the family members take place. In this regard, living space in the hotel (guestroom) may accommodate as a multifunctional space.
b. The guestroom should be flexible as to accommodate the prayer needs. The space should be flexible to allow the sajdah or prayer mat to be laid in the direction of the qiblah. Flexibility is one of the important characteristics of the Malay house. A minimum of two persons in congregational prayer is highly recommended prayers in congregation are highly encouraged in Islam.
c. As toilets in hotels are attached to the guestrooms, considerations should be given to the orientation and layout to be conducive for cleansing and performance of ablution or *wudhu'*. The toilets should provide bidet for wudhu and not have water closets in the direction of Qiblah.
d. Visual privacy at the guestrooms can take precedence to the design implemented in the Malay traditional houses. The control of the openings can be achieved through the layers and versatility of the opening design to ensure the privacy of the guests.

- Recreational Facilities

The design of recreational facilities in the hotel; i.e. swimming pool, spa, fitness center, and etc., should consider gender separation. The segregation applies most to female guests who require safety and security when enjoying the stay in hotel or resort. The layout of Malay house expresses clearly the division of space between male and female.

- Others

Bed and toilet planning in Malay house is not laid in the qibla direction. Therefore, this concept of orientation must be applied in the internal layout, i.e., bed arrangement.

A primary concern in Malay Muslim-friendly and SC hospitality building can be achieved through architectural design. The Malay image can be applied to the façade, spatial design, and interior decorations. However, to be classified as SC Hospitality facilities, the building should accommodate requirements of Islamic principles in the whole hospitality system; operation, design, and financial management. In other words, the design of the Islamic hospitality building should be the product of the Islamic beliefs and the value of the occupants.

CONCLUSION

The study concludes that there are profound vernacular features encompassing socio-cultural and environmental values, which compliments Syariah. Currently, the application of SC design in hospitality building needs to be outlined since there are no direct guidelines from the Quran and Sunnah. This paper has reviewed and outlined the pertinent issues to be adhered to in the provisions of SC hotel. The analysis conducted also indicated the cultural values and meaning of spaces of the Malay traditional houses that can be assimilated in the design of hospitality buildings. In addition, it is expected that implementing the essence of Malay identity in the hospitality building design will enhance the tourism industry as tourists will experience and learn more about the Malay culture. The results of the study outlined the cultural meaning of hospitality building that is successfully integrated into Syariah Compliant design. The findings are invaluable to reinstate the Malay identity into the contemporary hospitality building to draw higher numbers of visitors interested in cultural tourism. The analysis reveals that the concept of privacy and segregation are the main concern in the traditional Malay houses. These Islamic principles may be applicable to any hospitality building. On the other hand, looking retrospectively at the significant importance of Islamic cultural tourism, the hotel designs must embody cultural and traditional attractions. The hotel design can also integrate Islamic values to ease the Muslim to practice the requirements of Islam. There is still a need to consider an explicit understanding of the essence of Malay Muslim houses. The knowledge is vital to revitalize these values during the design of future hotels to enhance the Islamic cultural tourism industry.

ACKNOWLEDGEMENTS

The authors wish to thank the group leader and all members of NRGS research titledRating Tools and Audit Systems for Syariah Compliant Hospitality and Services for the fund and support.

REFERENCES

1. Hoseini.A. G., Baharuddin, M. N. & Ibrahim, R.(2009), Identification of the Malay vernacular spatial features within Malay vernacular houses.*International Journal of the Arts in Society*, 4(4), 137-152.

2. Kamaruddin, R., Iberahim, H. & Shabudin, A. (2013). Factors Influencing Customers Willingness to Pay for Halal Logistics. *Journal Of Asian Behavioural Studies*, 3(9), 41-52.

3. Lim, J. Y. (1987). *The Malay house: rediscovering Malaysia's indegenous shelter system*. Pulau Pinang: Institut Masyarakat.

4. Heritage Studies (2010). *Rumah Meor the legacy of traditional Perak house*. Department of Architecture. Kulliyah of Architecture and Environmental Design, IIUM. Unpublished report.

5. Mortada, H. (2003). *Traditional Islamic Principles of Built Environment*. Routledge Curzon: New York.

6. Muhammad, Z. (2009). Al Jawhara: Promoting Islamic hospitality and Halal tourism. *The Halal Journal*(May/June), 60.

7. Gurupiah, M. and Mohammad Tajuddin, R. (1998). The Basic Malay House and The Malay House of the peninsular west coast, In *Chen Voon Fee (ed.), The Encyclopedia of Malaysia Architecture (Volume 5).* Archipelago Press: Kuala Lumpur, Malaysia.

8. Abdul Halim, N. (1983). *Rumah kutai: Satu warisan senibina tertua di Perak*. Persatuan Muzium Malaysia, Muzium Negara: Kuala Lumpur. Abdul Halim, N. and Wan Hashim,W.T.(1997). *The Traditional Malay House*. Shah Alam: Fajar Bakti.

9. Noorul Huda, M. R., Anuar, T.(2013). The Concept of Privacy and The Malay Dwelling Interior Space Planning.*AMER International Conference on Quality of Life*. Procedia-Social and Behavioral Sciences 101(2013), 404-414.

10. Nur'Hidayah, C. A., Ahmad Hidaya,t A. R. & Mohd Salehuddin, M. Z. (2012). Dry Hotel and Syariah Compliant Practices: Concepts, Challenges and Reality in Malaysia.*International Conference on Innovation, Management and Technology Research (ICIMTR2012),* Malacca, Malaysia,107-111.

11. Sani, R. M. and Mahasti, Payam (2013). Sustainable Tourism Design through Preserving Regional Identity: Iran.*Journal of Asian Behavioural Studies,* 3(8), 101-113.

12. Zulkifli, Ibrahim, Mohd Salehuddin, Zahari, Maimunah, Sulaiman, Zulhan, Othman, Kamaruzaman, Jusoff (2009). Travelling pattern and preferences of the Arab Tourist in Malaysian Hotels. *International Journal of Business and Management.* 4(7). retrieved from www.ccsenet.org/journal.html

Chapter 10

The Role of Halal Awareness, Halal Certification, and Marketing Components in Determining Halal Purchase Intention Among Non-Muslims in Malaysia: A Structural Equation Modeling Approach

Yuhanis Abdul Aziz[a*] & Nyen Vui Chok[a]

ABSTRACT

Although the important role of halal awareness, halal certification, and marketing-related components in determining purchase intention has been recognized in the marketing literature, empirical examination of this issue is limited. In order to fill this knowledge gap, this study aimed to determine the relationships between halal awareness, halal certification, food quality, marketing promotion, and brand with the intention to purchase halal product among the non-Muslim community in Malaysia. The data were collected through a self-administered questionnaire survey consisting of 226 non-Muslim consumers. Structural equation analyses indicated that the halal awareness, halal certification, marketing promotion, and brand were positively related to purchase intention, whereas food quality was negatively associated with it. Theoretically, this study is one of the first attempts to develop and to empirically test a conceptual model on halal purchase intention by integrating halal components with marketing-related components.

Keywords: halal awareness, halal certification, Malaysia, purchase intention

INTRODUCTION

The halal market is estimated to be worth US$580 billion a year globally and the halal food industries estimated growth rate is 7% annually. According to Burgmann (2007), the halal growth status can be linked to religion and beliefs that it is cleaner, healthier, and tastier. The halal logo becomes a representation of quality measurement and religious fulfillment. Burgmann further stated that not only is halal subjected to the food segment but also halal denotation is expended to other nonfood segments. Previous research on halal food consumption has been done in some countries such as Belgium, France, and Saudi Arabia. In general, foreign foods in these countries have become assimilated and their original tastes are changing according to the preferable local taste, thus encouraged by tourism factors. Kebabs are characterized as German staples, whereas curry is the famous takeout meal in the United Kingdom.

Integration, advancement of technology, and continued improvement of activities has simplified the process during halal production. Manufacturing has since improved its processes effectively and efficiently. The globalization and advancement in science technology may contribute to mass production and enhancing the quality of production as well. Food and nonfood products categories have to fulfill the halal concept because the Muslims especially are becoming more halal conscious. They will distinguish a halal logo before making any purchase. Ingredients for the particular products need to be listed in detail because this acts as a reference for the nonhalal items.

Malaysia is one of the countries in the world where the government provides full support in promoting the halal certification process on products and services. Subsequent to the halal certification, Malaysia has developed halal certification as the total quality health and sanitary system in adopting procedure for slaughtering process and other related operations as prescribed by Islamic jurisdictions. The certification is not limited to the poultry and meat only; it also covers consumers' items such as cosmetics, pharmaceuticals, and toiletries. Food manufacturers and producers are required to comply with the benchmark standards of Hazard Analysis Critical Control Point (HACCP), Good Manufacturing Practices (GMP), Good Hygienic Practice (GHP), and ISO9000 in order to meet halal requirement.

Accordingly, Malaysia food and beverage manufacturers are aggressively promoting the industry to become the halal hub due to the global market opportunities. Sumali (2006) stated that manufacturers and producers who manage to obtain the halal products certification by Jabatan Kemajuan Islam Malaysia (JAKIM) are providing the consumers assurance that ingredients use the processing, preparation, hygienic, and cleanliness procedures complying with the halal requirement and are consistent with HACCP and other quality assurance standards.

To date, despite the widely available halal food and numerous research reports on the halal food market, there is a scarcity of theory-driven research on halal food purchasing (Alam & Sayuti, 2011). Moreover, there has been a lack of genuine knowledge about the relationship between halal concept such as halal awareness and halal certification with purchase intention in the context of purchasing halal product. Similarly, the knowledge about the impact of marketing-related concepts such as promotion and brands with halal purchase intention is almost unavailable. Though marketers are beginning to gain an understanding of the importance of halal certification and halal awareness, very little is known about halal components and other determinants that impact purchase intention. Furthermore, consumers are not exposed enough to halal certification and halal brands (Rajagopal, Ramanan, Visvanathan, Satapathy, 2011), and their level of awareness toward both halal concepts are relatively low. Thus, it is important that an understanding of why and how consumers perceive halal foods be explored further because it may help food manufacturers to market their products effectively. Additionally, due to the growing number of non-Muslim population in Malaysia, the study of halal food needs to be further examined among non-Muslims, especially in the context of recognition toward halal food. Halal products should be recognized as a symbol of cleanliness, safety, and of high quality, which may be cultivated among the non-Muslim consumers in Malaysia. To facilitate the progressive demand and increase the purchase halal food among non-Muslim consumers, it is therefore timely for the study to consider the factors influencing the intention of purchase. Moreover, the utilization of halal food such as meat consumption decisions within a religious context could differ significantly from purchase situations where religion does not play a key role (Kaynak, 2009). In-depth understanding of non-Muslim consumers' purchase behavior can strengthen the position in the halal marketplace.

By understanding the relationships between future behavioral intentions and their determinants, halal manufacturers and marketers' managers preferably should know how to build up an attractive image and improve their marketing efforts in order to maximize their sales and increase their market shares. Hence, the purpose of the study is twofold. The first is to develop and test a general model of customer purchase behavior in the context of halal study. The second is to examine the relationships between halal components, namely, the halal awareness and halal certification and marketing determinant (i.e., food quality, marketing promotion, and brand) in their prediction of future behavioral intentions. We believe that our proposed model can act as a crucial preliminary step toward a comprehensive understanding of consumer behavior in the context of intention to purchase for halal product.

The article is structured as follows: theoretical framework based on the theory of planned behavior is briefly reviewed, focusing on halal awareness and halal certification and how these components are related to the theory of planned behavior. The structural model depicting the proposed relationship and hypotheses formulation are then explained in detail. The method and data collection are then summarized. Results are presented next, beginning with a measurement model, followed by structural model and moving on to hypotheses testing. Finally, discussion on findings, limitations, implications, and future research direction are discussed.

THEORETICAL FRAMEWORK

The Theory of Planned Behavior (TPB)

This study uses the theory of planned behavior (TPB; Ajzen, 1991) as a basic underpinning theory to support the relationship between halal concept and other antecedents with intention to purchase by customers. TPB has been widely recognized and applied in food-related studies (Sparks & Shepherd, 1992; Verbeke & Vackier, 2004) and halal food study (Bonne & Verbekem, 2008). Similarly, previous studies on halal purchase intention apply the TPB as its theoretical foundation (see, e.g., Alam & Sayuti, 2011; Aziz, Amin, & Isa, 2010). According to TPB, there are three major components that influence human attitude to behave. The components are attitude, subjective norms, and behavioral control. TPB provides guidelines for predicting human social behavior.

In the context of this study, it is predicted that humans behave according to a predicted framework as suggested by Ajzen (1991) in purchasing halal food. The proposed framework for the study may not translate the TPB directly but it helps to explain the way in which individual human beliefs or perceptions are determined or influenced by the individual's wish to perform that behavior. The most relevant component of TPB is attitude toward behavior. Subsequently, attitude toward behavior, which originated from a collective behavioral belief such as belief that the product is halal may lead to favorable attitudes such as having intention to purchase. Subjective norm is, another function of belief, which is perceived social pressure to engage or not to engage in that particular behavior. However, in the context of this study subjective norms as well as perceived behavioral control are less relevant because this study integrates some of the basic assumptions of marketing by examining the relationships among three key variables that are deemed to be important in understanding halal concept related to purchase intention.

Malaysia is a country that consists of multiracial and multireligious communities where approximately 62% of the populations are Muslim and the remaining 38% are non-Muslim. Without a doubt, the halal concept is an absolute key to consumption for the Muslims, but nowadays, the halal concept receives growing attention among non-Muslim consumers due to several reasons, such as how food is prepared in a hygienic and safe environment as well as the slaughtering process, which is performed according to humane animal treatment. Hence, similar to the Muslim community, non-Muslims perceive halal as the standard of choice. The availability of halal food around the world and the growing numbers of non-Muslim halal food manufacturers has provided an indication that the demand for halal products is strong and increasing every year.

Therefore, the study of halal concepts among non-Muslims is critically important as marketing the halal product can provide market expansion and growth to food manufacturers especially in the non-Muslim market. As non-Muslims are able to accept halal products, the factors that influence the non-Muslim to purchase halal products demand further research. In particular, determining the attitude of non-Muslims and their behavior toward halal foods is vital when marketing halal products, especially in the context of intention to purchase.

Research on halal is still in its infancy. As previously mentioned, former studies on halal use the TPB and theory of reason action (TRA) as their foundation to understand customer behavior (Bonne, Vermeir, & Verbeken, 2009; Lada, Tanakinjal & Amin, 2009) toward halal perception. To date there is very limited research available to examine the relationship between halal issues with purchase intention. Bonne et al. (2009), for example, investigated the determinant of halal meat consumption among the Muslim community in Belgium. A study conducted by Luthfi and Imam (2010) took a look at the intention to switch from products without certified halal labels within a wide array of purchase context, especially in the context of food and medicine products. A recent study by Shaari and Arifin (2010) only looked at the halal purchase intention component in general without empirically examining the impact of each component on purchase intention. Thus, the causal relationships between halal components with intention to purchase have been only conceptually discussed. On the other hand, research done by Golnaz, Zainalabidin, Mad Nasir, and Eddie Chiew (2010) examined the relationship between halal concept and perceived behavioral control with behavioral intention to purchase halal products among the non-Muslim community. Although many studies in the halal field have dealt with purchase intention, relatively few studies have examined the role of promotion, product quality, brand, and halal issue in relation to purchase intention. This study offers an integrated approach to understanding customer behavior and extends the theoretical and empirical evidence on the causal relationship between the proposed constructs. Accordingly, a research model is proposed and tested using a structural equation model in the study. It is hoped that the findings derived from the model will serve as the basis for the development of halal marketing strategies, especially in the context of halal food.

THE STRUCTURAL MODEL AND HYPOTHESES

This article develops a structural model as shown in Figure 1. The purpose of the framework is to examine the relationship between halal awareness, halal certification, food product quality, marketing promotion, and brand with intention to purchase halal products. Each component of the model was selected on the basis of the literature review. Accordingly, six constructs were conceptualized to fit into the current study setting.

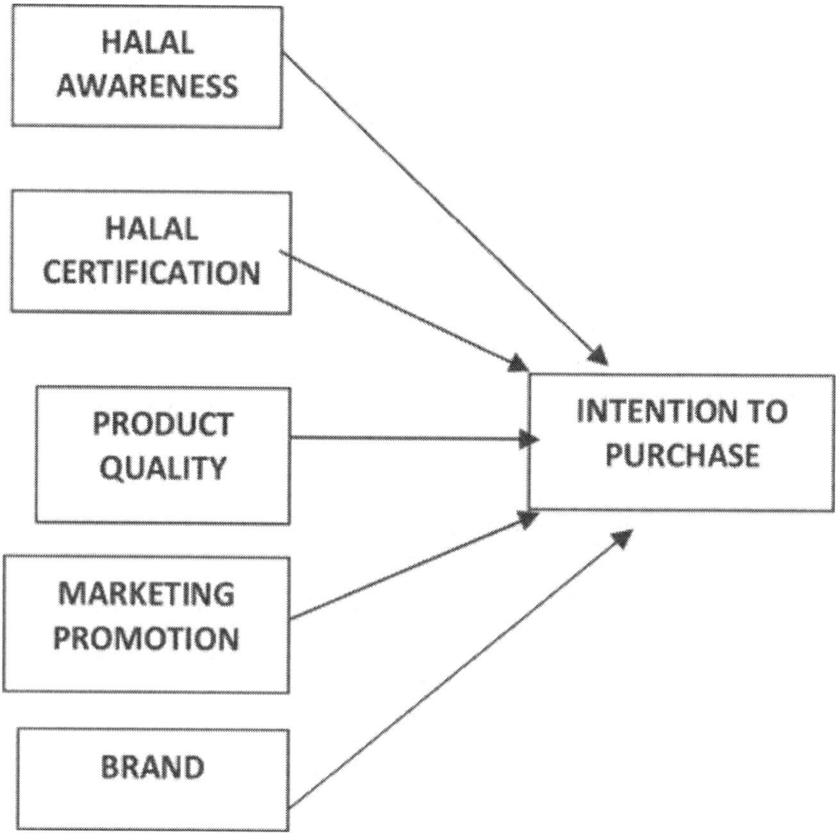

Figure 1 Proposed structural model.

Halal awareness is conceptualized to correlate with intention to purchase. It adds to the degree of the consumer's knowledge on halal foods. Similarly, halal certification is added to the model. It is critical to include halal certification because non-Muslim consumers compared with Muslims view it differently. To the non-Muslim consumer, food that is certified halal is tastier, hygienic, and safer. In the context of marketing determinant, food quality is proposed to have an influence on intention to purchase. Previous research on quality and purchase intention and behavior relationship showed that there was a significant impact of buying decision between products' qualities. A better quality product would lead to stronger purchase intention for customers (De Cannière, Pelmacker, & Geuens, 2009). Next, marketing promotion is included as it is important in terms of influencing customers to purchase the product. It is equally important to understand and analyze consumers' brand choice. Hence, brand is added as another important marketing-related construct in the model. Brand as an image and reputable status for products (halal product) could increase consumer confidence in the general perception of the product. The theoretical underpinning of the model is discussed in the following section.

Halal Awareness

Awareness is the ability to perceive, to feel, and to be conscious of events and objects. It is a concept about implying the understanding and perception toward the events or subjects. Awareness has been hypothesized as an important role in determining the intention to choose. Golnaz et al. (2010) found in their study that the awareness of halal principles and halal food products is determined by a positive attitude. Consonant with the TPB theory, consumers who have a favorable attitude will perform the behavior (which in the context of the current study is behavior to consume or purchase halal product). Empirical evidence provided by Lada et al. (2009) in their study on halal product confirmed that attitude is positively related to intention to choose halal products. Alam and Sayuti (2011) also reported similar results for a sample of marketing students in one of the universities in Malaysia. Hence, Lada et al.'s findings have verified that decision to choose a halal product is determined by a positive attitude. In the context of the current study, the positive attitude is the favorable perception of halal concept and halal awareness. Accordingly, the following hypothesis is proposed:

H1: Halal awareness is positively related to intention to purchase.

Halal Certification

Given the large number of certification systems in the food industry, it is surprising that there are only a few research approaches to the economics of certification. Halal certification refers to the official recognition of the orderly process of preparation, slaughtering, cleaning, handling, and other relevant management practices by the established body (such as JAKIM in Malaysia). In order for food to be certified halal, the manufacturer must acquire the halal symbol or halal qualifications as evidence that the products are religiously lawful according to holy Quran (Guntalee & Unahannda, 2005). Certified halal food is a requirement for the Muslim as part of religious obligation. Interestingly, the concept of halal is more than what it used to be due to its wide acceptance by both Muslims and the non-Muslims. Halal concept is deliberated as the standard of choice for these two groups worldwide (Golnaz et al., 2010). Accordingly, certified halal food may also signal that the food adheres to stringent standards in hygiene and sanitation (Lada et al., 2009). There is empirical evidence to support the premise that non-Muslims are concerned about food safety, which positively influences the probability of their attitude on halal product (Golnaz et al., 2010). Moreover, the finding of their study has suggested that attitude toward halal food and perceived control are significant predictors of intention. Hence, in line with TPB theory (Ajzen, 1985, 1991), the perception of halal food in the context of halal certification by the non-Muslim is determined by a positive personal attitude (Golnaz et al., 2010), which in turn may influence their intention to purchase halal food. The argument has led to the formation of the following hypothesis:

H2: Halal certification is positively related to intention to purchase.

Food Quality

Food quality or product quality refers to physical product attributes such as taste, appearance, and other attributes. Quality is a desirable characteristic of a product or service that is demanded by the customers (Canavari, Castellini, & Spadoni, 2010). Food quality can be a source of product differentiation and is considered a key parameter for the food industry as it is the basis for success in today's dynamic and highly competitive market (Du & Sun, 2006). Hence, understanding the relationship between food quality and customer behavior is important so that the manufacturer can remain competitive in the market. One of the critical areas that need's further investigation is the impact of food quality on customers' intention to purchase because purchase intentions are commonly used as a basis to forecast purchase behavior. The knowledge of this relationship is important to food manufacturers because it provides basic information on how to meet the demands of the customers so that they can be satisfied. In this study, food quality is perceived from two broad contexts. First, within the halal context, the food is considered to possess quality if it meets several requirements such as safe to consume, healthy, and hygienic. The second perspective is within the general context of food quality, which covers areas such as the food that is offered is superior to the competing product and the food product matches the consumer's ideal product. Food quality can be a source of product differentiation to a food manufacturer. According to Newberry, Klemz, and Boshoff (2003), food quality is regarded as one parameter in predicting purchase behavior. Therefore, based on this argument the following hypothesis is constructed:

H3: Food quality is positively related to intention to purchase.

Marketing Promotion

Promotion, also known as marketing communication, is one of the four Ps in marketing mix. According to Grönroos (1994), marketing mix is able to influence a firm's competitive position. Promotion is a useful marketing tool (Kotler & Armstrong, 2006), which may influence purchases among consumers. For the purpose of the current study, food promotion is defined as marketing and sales promotions used on halal food labels or as halal food packaging designed to entice consumers to buy a product at the point-of-sale (Hawkes, 2004). Promotion is an important marketing tool used to entice consumers to buy a product or service at the point-of-sale. Promotions may consist of premium offers, such as giveaways and competitions; celebrity endorsements; and cartoon and movie character promotions (Hawkes, 2004). Essentially, marketing promotion has a specific role in creating brand awareness, generate favorable brand attitude, and stimulate purchase intention (Belch & Belch, 1998; Rossiter & Percy, 1998). Hence, the following hypothesis is forwarded:

H4: Marketing promotion is positively related to intention to purchase.

Brand

According to Dodds and Monroe (1985), brand name has extrinsic quality cues. Han (1989) views brand name as a summary construct for quality because it has inference quality based in brand name. Kotler and Amstrong (2006) have acknowledged that a particular brand might not only be represented by a name or symbols. It represents consumers' perception and sentiment toward the product and service, which means to the consumers' point of view. Brand name can affect consumers' preferences and intention to purchase (Alreck & Settle, 1999; Ataman & Ulengin, 2003). Past research has suggested that customers' intention to purchase a product or service can be influenced by positive attitude toward brand itself (Laroche & Brisoux, 1989), Consumers would choose a brand that is similar to its characteristics. In line with the TPB theory, customer's who have a positive attitude toward brand, for example, agree to choose a particular food brand are able to act further by demonstrating positive effect on halal purchase intention. Therefore, a hypothesis was conceptualized as follows:

H5: Brand is positively related to intention to purchase.

METHOD AND DATA COLLECTION

The data used in this study were collected via self-administered questionnaires from customers and the study was conducted in Lembah Kelang, Selangor, Malaysia. Respondents were approached randomly on streets and around shopping malls to participate in the survey. Malaysia as a multiracial country with over 38% of non-Muslim population has provided a landscape for this study. Empirical research using survey was carried out to examine the proposed structural model and suggested hypotheses. Each of these constructs was measured using established, close-ended questions, which were conveniently distributed to non-Muslim respondents. Descriptive statistics, reliabilities, and correlations using SPSS 18 were computed for each construct to obtain the results and to identify the relationships. The structural equation modeling approach using a two-step process was applied in this research.

Measure

The instruments used to measure each of the constructs were multi-item scales, which were adopted from previous established research, which were then modified to fit the context of this study on the purchase intention of non-Muslims toward halal products. The validity and reliability of the items are proven and have been applied considerably by previous researchers. The questions were prepared using 7-point Likert scale questionnaires. The seven anchors that were used in the questionnaires were (1) *strongly disagree*, (2) *disagree*, (3) *somewhat disagree*, (4) *neutral*, (5) *somewhat agree*, (6) *agree*, and (7) *strongly agree*. The questionnaires were later subjected to a pretest because without the pretest, the researcher is unable to determine the degree of acceptance of questionnaires among respondents (Bell, 2005). The questionnaires were evaluated by a group of experts that consisted of academic staff and practitioners in halal-related business. A pretest was administered at a major shopping mall in Klang Valley on 30 respondents. In this study, the questionnaire is divided into two sections; the first section covers the dimensions toward purchase of halal products, whereas the second section is about demographic question. The Appendix lists the descriptions of measurement of constructs for the study.

Sample

A total of 226 respondents participated in the study. There were more female (64.6%) than male (35.4%) respondents. The majority of respondents were Chinese (71.24%) followed by Indian (26.11%) and other respondents (2.65%) of Iban descent and other Bumiputera groups from Sabah and Sarawak (West Malaysia). The highest percentage fell under the age group of 20 to 29 years old (72%). Respondents were mostly working in the private sector (48.2%), students (27.9%), and government servants (21.1%). Table 1 provides a demographic profile of respondents.

Table 1 Demographic Profile of Respondents

Characteristics	%
Gender	
Male	35.40
Female	64.60
Race	
Chinese	71.24
Indian	26.11
Other	2.65
Age	
20–29	72.12
30–39	22.57
40–49	4.42
50–59	0.88
Marital status	
Single	76.11
Married	22.57
Other	1.33
Occupation	
Government	21.24
Private sector	48.23
Student	27.88
Unemployed	2.21
Other	0.44
Level of education	
SPM and below	1.77
STPM	3.10
Diploma/Certificate	16.37
Degree	57.08
Master/Doctorate	20.35
Other	1.33
Monthly income	
Below RM 1,500	29.65
RM 1,501 - RM 3,000	37.17
RM 3,001 - RM 4,500	20.80
RM 4,501 - RM 6,000	8.85
Above RM 6,000	3.54
Level of religiosity	
Not at all	20.80
Not devout	6.64
Fairly not devout	20.80
Fairly devout	31.86
Devout	15.93
Extremely devout	3.98

DATA ANALYSIS AND RESULTS

Structural equation modeling (SEM) using AMOS 16 was applied to simultaneously examine the model and hypothesized relationships proposed in the study. The current study adopted the most widely used two-step techniques recommended by Anderson and Gerbing, (1988). One of the requirements is that the measurement model that measures individual latent construct using multiple indicators needs to be analyzed using confirmatory factor analysis (CFA) so that its validity can be confirmed. CFA of the measurement model is considered appropriate when there is a theoretical and empirical knowledge of the underlying latent variable structure (Anderson & Gerbing, 1988; Byrne, 2001). At this stage, the strength of the regression paths from constructs to their observed variables is the key interest of the research. The primary concern is the extent to which the observed variables are generated by the underlying constructs. Once the measurement model is validated and goodness-of-fit is satisfied the model is ready for the next analysis, that is, structural model.

Measurement Model Testing

The results of means, standard deviation, Cronbach's alpha, and correlations among the variables are shown in Table 2. As demonstrated in Table 2, the scale alpha reliabilities for all constructs are above. 70 indicating the reliability of each construct except for promotion construct.

Table 2 Summary of Mean, Standard Deviation, and Correlations Among the Constructs (N = 226)

Variables	M	SD	Cronbach's Alpha	1	2	3	4	5	6
1. Halal awareness	3.62	0.804	.789	1					
2. Halal certification	3.30	0.934	.836	.621**	1				
3. Food quality	3.12	1.122	.821	.267**	.375**	1			
4. Promotion	3.62	0.731	.673	.298**	.245**	.020	1		
5. Brand	3.26	1.033	.849	.399**	.426**	.184**	.366**	1	
6. Intention to purchase	3.41	0.876	.824	.538**	.651**	.258**	.355**	.560**	1

**Correlations are significant at 0.01 levels (two-tailed).

CFA analysis was conducted to examine the validity of the measurement model. The overall fit of the CFA as well as structural models were examined using common parameters, namely, chi-square statistics, goodness-of-fit index (GFI), comparative fit index (CFI), and root mean square error of approximation

(RMSEA; Hair, Black, Babin, Anderson, & Tatham, 2006). As a rule of thumb, the recommended cutoff value for GFI, NFI, and CFI is ≥ .90, and the acceptable threshold level for RMSEA is ≤ .08 (Hu & Bentler, 1998). The model fit the data relatively well (see Table 3). According to the goodness of-fit indices (χ^2 = 326.887, *df* = 137, RMSEA =.078, GFI = .869, CFI = .939, NFI =.901) from CFA, the measurement model provided evidence of satisfactory fit. Hence, the fit measure of the model suggested a reasonable fit. Simultaneously, the validity test that covers convergent and discriminant validity was examined. Convergent validity was assessed by examining construct loadings and construct reliability (Peter, 1981). It was found that all construct loadings were significant ($p < .001$) and above the recommended value, that is, 0.7. Average variance extracted (AVE) was also employed to assess convergent and discriminant validity. As shown in Table 4, all AVE exceeded the recommended cutoff point of .5 (Bagozzi & Yi, 1998) suggesting convergent validity was satisfied. Also, an AVE for each construct was greater than squared correlation coefficients for corresponding interconstructs, which confirms discriminant validity (Fornell & Larcker, 1981). On the other hand, composite reliability was evaluated using the Fornell and Larcker (1981) formula. As shown in Table 4, the reliability indices were above the recommended thresholds of .60. Standardized loadings, AVE, and composite reliability are shown in Table 4.

Structural Model Testing

After the overall measurement model was found to be acceptable, the structural equation model was conducted to examine the relationship among the studied constructs. It was tested with the entire sample (N = 226). It was specified based on the hypotheses proposed. The results of SEM analysis are depicted in Figure 2. The fit indices of the model are summarized in Table 3. The initial structural model suggested that the data fits the model reasonably well. However, one of the indices, that is, GFI, suggested that the model was modified further in order to improve goodness-of-fit. Hence model modification was necessary. Modification indices suggested that Item PI 1 (Choosing halal products is a good idea) needs to be deleted due to relatively low standardized regression weight (.603) compared with other items and high error variance (.734). Thus, it is reasonable to rationalize that deletion of this item is necessary to improve the goodness-of-fit of the model. Item PI 1 was therefore deleted from further analysis. The result of the modified model demonstrates a satisfactory model with adequate goodness-of-fit (χ^2 = 195.222, *df* = 103, RMSEA = .063, GFI = .911, CFI = .967, NFI = .933). Accordingly, the result suggests that the hypothesized model fits the empirical data well.

Table 3 The Empirical Result of Halal Model With Parameter Estimate

Model	χ^2	*df*	χ^2/df	RMSEA	GFI	NFI	CFI
Measurement model	326.887	137	2.386	.078	.869	.901	.939
Initial structural model	296.160	136	2.170	.072	.880	.910	.949
Final structural model	195.222	103	1.895	.063	.911	.933	.967

Table 4 Summary of Factor Loading, Average Variance Extracted, and Composite Reliability

Factor/Items	Standardized Factor Loading	Average Variance Extracted	Composite Reliability
Halal awareness			
I will make sure the product has gone through the slaughtering method before purchasing it	.764	0.73	0.97
I will only purchase the product if I know the slaughtering process	.885		
I will only purchase the product if the slaughtering process follows the Islamic rules	.825		
Halal certification			
The halal logo is important in choosing the product	.766	0.66	0.96
I will choose the product based on the halal logo	.755		
I will always be careful when choosing products with the halal logo	.805		
I am aware of the differentiation between the genuine and nongenuine halal logo	.739		
Food quality			
Offers advantages that are not offered by competing products	.880	0.83	0.95
Marketing promotion			
I will purchase halal food during sales promotion	.810	0.78	0.93
I will purchase halal food when discounts are given	.846		
Brand			
The halal brand is high quality	.791	0.69	0.96
The halal brand is widely popular	.845		
The halal brand is a reliable product	.852		
Intention to purchase			
Choosing halal products is a good idea	.687	0.68	0.98
My friends also think that I should choose halal foods	.827		
Most people who are important to me choose halal foods	.865		
My family members prefer halal foods	.862		
I like to choose halal foods	.899		
I will recommend friends to purchase	.822		

Figure 2 Final structural model.

Hypotheses Testing

The proposed hypotheses were tested using a series of simultaneous regression analysis using SEM with maximum likelihood estimation. Within the overall model, the estimates of the structural coefficients provide the basis for testing the proposed hypotheses. Hypotheses were tested by examining the significance level, direction, and magnitude of the standardized estimates of paths that link independent variables with the dependent variable. Results are summarized in Table 5. All hypothesized relationships were significant. 001 except for the halal awareness → intention to purchase and promotion → intention to purchase path, which were significant at. 05.

Table 5 Results of Hypotheses Testing

Hypothesis	Estimate	p Value	Result
H1: Halal awareness → Intention to purchase	.142	< .005*	Supported
H2: Halal certification → Intention to purchase	.302	< .001	Supported
H3: Food quality→ Intention to purchase	.056	< .001	Not supported
H4: Promotion → Intention to purchase	.150	< .005*	Supported
H5: Brand→ Intention to purchase	.323	< .001	Supported

*p value <.005.

The directional relationships between halal awareness and intention to purchase, halal certification and intention to purchase, promotion and intention to purchase, and brand and intention to purchase were statistically established (as shown in Table 5 and Figure 2). The findings therefore indicate support for a positive relationship for H1, H2, H4, and H5. As a result, these hypotheses are supported. However, as shown in Table 5, food quality has shown no significant relationship with intention to purchase. Due to insignificance in the standardized regression weight, H3 is therefore not supported.

DISCUSSION

The focus of this study was on investigating customer behaviors pertaining to their purchase intention toward halal product. Using the TPB as the landscape in understanding customers' attitude and behavior toward halal purchase, the study develops a structural model by constructing halal elements (i.e., halal awareness, halal certification) and marketing components (i.e., food quality, marketing promotion, and brand) to be evaluated against intention to purchase. The structural relationships between all variables in the study were tested using the data obtained from a consumer questionnaire survey in Malaysia. The results provide a number of useful insights into the importance of halal concept in predicting consumer purchase intention. Of the five hypotheses, four were duly supported.

The structural relationship analysis indicates that halal awareness appears to have an effect on purchase intentions (i.e., intention to purchase halal product). The result indicates that halal awareness demonstrated by non-Muslim consumers had positive effects on intention to purchase halal product. It could be argued that people's attitude toward intention to purchase is determined by their level of awareness toward the concept of halal products. This is congruent with the TPB theory (Ajzen, 1991) and other empirical study on halal food purchasing (Shah Alam & Sayuti, 2001). Accordingly, attitude in the context of awareness toward believing in halal product appear to suggest direct relationship with intention behavior as postulated by TPB. This finding is consistent with Golnaz et al. (2010).

The study also confirmed that halal certification has a significant positive relationship with intention to purchase halal product. It implies that halal certification provides recognition and quality assurance to non-Muslim consumers when they display their willingness to purchase in the future. The findings are in accordance to Hassan's (2009) argument where she claimed that certified halal product is not only welcomed by Muslim consumers but is also accepted by non-Muslim consumers. Accordingly, halal certification is not just a religious requirement for the Muslims but also a prerequisite for non-Muslims in choosing their food. Such findings are in accordance with the TPB theory. These findings put forward that attitude toward the judgment on halal certification has successfully encouraged behavior as a feature that influences intention to purchase. A product with halal certification is deemed fit for Muslim consumers because it fulfills the requirements of Sharia law (Jabatan Kemajuan Islam Malaysia [JAKIM], 2009) and it represents a symbol of quality (Hassan, 2009) However, in the context of current study, it is interesting to see that the non-Muslim community considers halal certification significant on their intention to purchase halal product. This suggests that the certification of halal product has provided a strong recognition of the quality of the food product, which contributes to healthier living style for the non-Muslim. Subsequently, the current study has provided a strong support for TPB, similar to other food-related studies (Cook, Kerr, & Moore, 2002; Sparks, Shepherds, Wieringa, & Zimmermanns, 1995). Hence, halal certification does play an important role in affecting future behavioral intentions of customers.

Surprisingly, empirical results reported insignificant relationship between food quality and intention to purchase halal food. This disappointing finding is therefore contrary to the findings by Newberry et al. (2003), where they found that food quality is the predictor of purchase intention. One possible explanation that we could propose is that the current research only tested general perception of food quality but did not test the individual quality component related to halal product. Therefore, the result seemingly suggests that when predicting food quality having an impact on purchase intention, it could be the quality of halal component in the product that is important rather than general perception of quality that matters. This could suggest that the elements of cleanliness, hygiene, safety, and suitable environment become the benchmark standards to evaluate food quality. Furthermore, the insignificant relationship could also be associated with the scale that is employed in the study. The scale used may not be able to capture the halal product quality as perceived by respondents. For example, the study does not treat the element of food safety as part of food-quality construct. Thus, the absence of the food-safety element in the food-quality measurement may reduce the strength of food-quality construct. This is important because food safety is considered the most basic aspect of food quality and an aspect that is prominently of public interest (Canavari et al., 2010).

The finding also demonstrates that promotion has significant influence on purchase intention. Thus, it is clear that promotion does influence a consumer's intention to purchase halal product. The result indicate that promotion plays an important role in persuading and motivating consumers to purchase the product. Our finding therefore confirmed the TPB where the effect of promotion on halal food has appeared to modify consumer attitude toward product, that is, attitude toward having intention to purchase. Another notable finding is that the current study found significant relationship between product brand and halal purchase intention. Thus, the study provides evidence that brand is an important determinant for halal purchase intention. It is clear that consumer perception on brand exerts influence on consumer purchase intention.

This study has several important contributions. Theoretically, this research is one of the first attempts to develop and empirically test a conceptual model on halal purchase intention by integrating halal components with marketing-related components. Previous studies on halal have focused on the examination of TPB application (Alam & Sayuti, 2011) or TRA (Lada et al., 2009) in halal context. Similarly, other studies had only concentrated on consumers' attitude toward halal product (Omar, Mohammad, & Omar, 2008) and consumer awareness and perceptions of halal concepts (Golnaz et al., 2010). The study of consumer behavior in the context of halal purchase intention is the starting point for any efforts attempted to understand consumer behavior, which is a dynamic and continuous process. These relationships might serve as a guide for researchers when designing their research. Studying consumer behavior is an integral part of successful strategic marketing because it enables marketers to understand and predict how consumers will act (Schiffman & Kanuk, 2004) . It is important for halal food manufacturers and marketers to understand how non-Muslims react toward halal concept and also their preferences on halal food. From this point of view, this study contributes to extant literature by offering a comprehensive halal purchase intention model in the course of consumer behavior that deliberately scrutinized and delineated the relationships between halal concepts and other marketing-related constructs. Hence, endeavors to build or improve halal awareness among the non-Muslims would facilitate their purchasing or recommending behaviors, thus being critical to the success of halal food manufacturers.

Another contribution was that the halal certification was more predictive of halal purchase intention than was halal awareness as reported in Table 5. This denotes that halal certification is regarded as a very crucial element for the consumer in the context of purchase intention. Halal certification and endorsement status has therefore provided robust evidence of the non-Muslim customers' acceptance of halal food as truly authentic, particularly in Malaysia. Accordingly, halal certification has boosted non-Muslim trust toward halal product due to its strict hygiene and other quality-related practices.

IMPLICATIONS

Our research conceived some implications. First of all, assurance of halal in terms of certification has signaled to the food manufacturers that obtaining halal certification must be considered a necessary element if they want to compete in the global arena. Hence, food manufactures who are still not obtaining halal certification must apply for the endorsement in order for them to extend the business and at the same time to be able to attract both Muslim and non-Muslim consumers. The findings clearly indicate that even though halal certification is an option for food manufacturers, they should consider applying for certification if they aim to capture the market globally. Moreover, in the context of Muslim consumerism, halal certification is a religious requirement. Thus, certification will assist in expanding the market locally and globally because the potential halal market is indeed enormous.

Second, the findings revealed that non-Muslim consumers reported significant levels of awareness toward halal product. It suffices to suggest that the non-Muslim community has adopted halal food as part of lifestyle choice. This means that they admitted the benefit of halal product, which imposes influences on their attitude and behavior to act in terms of intention to purchase the product. Accordingly, because they demonstrate their intention to purchase, this signifies that they are willing to pay for halal product. Clearly, this gives implication to halal food producers. Thus, it is critical for food manufacturers to increase the level of awareness toward halal product by providing sufficient and interesting information, especially on halal certification. Hence, manufacturers must take the opportunity to do intensive promotion in order to encourage more consumers to purchase their product. Similarly, it is critical for food manufacturers to maintain the reputation of their product brand because brand is another important predictor of halal purchase intention as reported in the study. The results imply that halal food manufacturers and marketers should significantly consider brand, food quality, and promotion when attempting to market their product because these marketing-related strategies may assist in boosting the halal sales and increase market shares.

LIMITATIONS AND FUTURE RESEARCH

Our study is not free from limitations and this provides us with suggestions for future research directions. Our study is limited in the fact this study did not investigate every possible marketing element that could affect or influence purchase intention besides marketing promotion. These may include, for instance, price and place. Hopefully, future research will incorporate these variables into research scope. This type of examination is important if the halal manufacturer or producer wants to understand more about customer preferences toward price related to halal and place suitability. Next, this study was constrained to respondents from a single country and from a single group, namely, non-Muslims, and this may well limit generalizations that may be made to the halal food industry in other countries. Therefore, future research should extend this study to a wider spectrum of Muslim community and to other parts of the world, particularly in the Western countries. It would be interesting to see if the Muslim community in a different country would react similarly or differently to the conceptualized model tested in the study. Additionally, it is important to test the model from the Muslim perspective because the consumption of halal meat for Muslims is quite different from the consumption of "regular" meat or other foods for non-Muslims (Bonne et al., 2009). Moreover, research in different settings may provide further validation of our tested model and it might also enable possible comparison with our findings.

Another limitation discovered can be associated with some measures used in the study. The significant relationship between food quality and purchase intention was not established, which could be mainly associated with the measures employed. Better measures involving specific facets that encapsulate halal food-quality features can be developed for this construct in future research. Finally, our work can be expanded with an investigation of food product quality with purchase intention. Although we have failed to establish its association, further research is needed in order to assess their relationship. In light of these considerations, it is hoped that the outcomes of this study will provide a platform for further research work.

CONCLUSION

Building on the TPB and marketing-related factors, this study empirically provided evidence of the significant relationships between halal awareness, halal certification, food quality, marketing promotion, and brand with halal purchase intention using data collected with non-Muslim consumers in Malaysia. One of the most important conclusions that may be drawn from this study lies in the fact that halal awareness, and halal certification are significant factors in explaining the intention to purchase halal product in non-Muslim societies. Additionally, important marketing-related factors play a significant role in predicting the intention to purchase halal product in non-Muslim societies.

Our tested model will be useful not only in halal academic research but also for halal manufacturers and marketing practitioners. The proposed model and conceptualization can act as a platform for the relevant practitioners and policymakers in managing the direction of halal industry development. Overall, our study has provided evidence of the existence of halal awareness among the non-Muslim community in Malaysia. Our study highlights that the TPB model is effective in examining the consumer's behavior purchase intention toward halal foods. Finally, this study has added to the current body of knowledge relating to general concept of halal food and consumer behavior in a developing multicultural country. It has the potential to be the basis of further explorations of halal food consumption within other countries and other societies.

CONTRIBUTORS

Yuhanis Abdul Aziz is a senior lecturer at the Department of Management and Marketing, Faculty of Economic and Management at Universiti Putra Malaysia.

Nyen Vui Chok is a PhD student at Graduate School of Management, Universiti Putra Malaysia.

REFERENCES

1. Ajzen , I. (1985). From intention to actions: A theory of planned behaviour . In J. Kuhl , & J. Beckmann , (Eds.), *Action control from cognition to behavior* (pp. 11 – 39). New York , NY : Springer .

2. Ajzen , I. (1991). The theory of planned behavior, organizational behavior and human . *Decision Processes* , 50 , 199 – 211 .

3. Alam , S. S. , & Sayuti , N. M. (2011). Applying the theory of planned behavior (TPB) in halal food purchasing . *International Journal of Commerce and Management* , 21 (1), 8 – 20 .

4. Alreck , P. L. , & Settle , R. B. (1999). Strategies for building brand preference . *Journal of Product & Brand Management* , 8 (2), 131 – 144 .

5. Anderson , J. C. , & Gerbing , D. W. (1988). Structural equation modeling in practice: A review and recommended two-step approach . *Psychological Bulletin* , 103 (3), 411 – 423 .

6. Ataman , B. , & Ulengin , B. (2003). A note on the effect of brand image on sales . *Journal of Product & Brand Management* , 12 (4), 237 – 250 .

7. Aziz , A. , Amin , M. , & Isa , Z. (2010). *The perception to choose halal cosmetics products: An empirical study for Malaysian consumers.* Retrived from http://icbme.yasar.edu.tr/previous_conferences/2010/e-proceeding/Azmi_AbdAziz.pdf .

8. Bagozzi , R. P. , & Yi , Y. (1988). On the evaluation of structural equation models . *Journal of the Academy of Marketing Science* , 16 , 74 – 94 .

9. Belch , G. E. , & Belch , M. A. (1998). *Advertising and promotion* (International ed.). New York , NY : Irwin, McGraw-Hill.

10. Bell , J. (2005). *Doing your research project ()* , 4th ed. . New York , NY : McGraw-Hill .

11. Bonne , K. , & Verbeken , W. (2008). Religious values informing halal meat production and the control and delivery of halal credence quality . *Agriculture and Human Values* , 25 , 35 – 47 .

12. Bonne , K. , Vermeir , I. , & Verbeken , W. (2009). Impact of religion on halal meat consumption decision making in Belgium . *Journal of International Food & Agribusiness Marketing* , 21 (1), 5 – 26 . [Taylor & Francis Online]

13. Burgmann , T. (2007 , July 22). Growing Muslim population pushing companies to produce products they can eat. The Star. Retrived from www.thestar.com/Business/article/238551.

14. Byrne , B. M. (2001). *Structural equation modeling with AMOS: Basic concepts, applications, and programming* . Mahwah , NJ : Erlbaum .

15. Canavari , M. , Castellini , A. , & Spadoni , R. (2010). Challenges in marketing quality food products . *Journal of International Food & Agribusiness Marketing* , 22 (3–4), 203 – 209 .

16. Cook , A. J. , Kerr , G. N. , & Moore , K. (2002). Attitudes and intentions towards purchasing GM food . *Journal of Economic Psychology* , 23 , 557 – 57 .

17. De Cannière , H. M. , Pelsmacker , P. D. , & Geuens , M. (2009). Relationship quality and the theory of planned behavior models of behavioral intentions and purchase behaviour . *Journal of Business Research* , 62 , 82 – 92 .

18. Dodds , W. B. , & Monroe , K. B. (1985). The effect of brand and price information on subjective product evaluations . In E. Hirchman & M. Holbrook (Eds.), *Advances in consumer research* , (Vol. 12 , pp. 85 – 90). Provo , UT : Association for Consumer Research .

19. Du , C.-J. , & Da-Wen Sun , D.-W. (2006). Learning techniques used in computer vision for food quality evaluation: A review . *Journal of Food Engineering* , 72 , 39 – 55 .

20. Fornell , C. , & Larcker , D. F. (1981). Evaluating structural equation models with unobservable variables and measurement error . *Journal of Marketing Research* , 18 (1), 39 – 50 .

21. Golnaz , R. , Zainalabidin , M. , Mad Nasir , S. , & Eddie Chiew , F. C. (2010). Non-Muslims' awareness of halal principles and related food products in Malaysia . *International Food Research Journal* , 17 , 667 – 674 .

22. Grönroos , C. (1994). From marketing mix to relationship marketing: Toward a paradigm shift in marketing . *Management Decision* , 32 (2), 4 – 32 .

23. . Guntalee , R. , & Unahannda , S. (2005). Needs, behaviour and attitudes of people in United Arab Emirates towards consuming Thai-Halal packaged food . *The Business Review* , 4 , 274 .

24. Hair , J. F. , Black , W. C. , Babin , B. J. , Anderson , R. E. , & Tatham , R. L. (2006). *Multivariate Data Analysis. ()* , 6th Ed. . Upper Saddle River , NJ : Pearson .

25. Han , C. M. (1989). Country image: Halo or summary construct? *Journal of Marketing Research* , 26 (2), 222 – 229 .

26. Hassan , W. M. (2009). *The growth of halal tourism and its implication for hospitality research* . In Y. A. Aziz , H. Hassan , W. M. Hassan , & M. S. Othman (Eds.), *Current issues in tourism and hospitality services in Malaysia* (pp. 85 – 102). Serdang , Malaysia : UPM .

27. Hawkes , C. (2004). *Marketing food to children: The global regulatory environment* . Geneva , Switzerland : World Health Organization.

28. Hu , L. , & Bentler , P. (1998). Evaluating model fit . In R. H. Hoyle (Ed.), *Structural equation modeling: Concepts, issues, and applications* (pp. 76 – 99). London , UK : Sage .

29. Jabatan Kemajuan Islam Malaysia (JAKIM). (2009). Retrived from http://www.halaljakim.gov.my/corporate

30. Kaynak , E. (2009). Editorial . *Journal of International Food & Agribusiness Marketing* , 21 , 1 – 4 .

31. Kotler , P. , & Armstrong , G. (2006). *Principles of marketing)* , (11th ed. . Upper Saddle River , NJ : Prentice Hall .

32. Lada , S. , Tanakinjal , H. G. , & Amin , H. (2009). Predicting intention to choose halal products using theory of reasoned action . *International Journal Islamic and Middle Eastern Finance and Management* , 2 (1), 66 – 76 .

33. Laroche , M. , & Brisoux , J. E. (1989). Incorporating competition into consumer behaviour models: The case of

attitude intention relationship . *Journal of Economic Psychology* , 10 , 343 – 362 .

34. Luthfi , B. A. , & Imam , S. (2010). Marketing impact of halal labeling toward Indonesian muslim consumer's behavioral intention based on Ajzen's planned behavior theory: Policy capturing studies on five different product categories. *Proceedings of the fifth International Conference on Business and Management Research.* Retrived from http://ssrn.com/abstract=1682342

35. Newberry , R. C. , Klemz , B. R. , & Boshoff , C. (2003). Managerial implications of predicting purchase behavior from purchase intentions: A retail patronage case study . *Journal of Services Marketing* , 17 (6), 609 – 620 .

36. Omar , W. M. , Muhammad , M. Z. , & Omar , A. C. (2008). *An analysis of the Muslim consumer's attitudes towards "halal" food products in Kelantan.* Paper presented at ECER Regional Conference, Kelantan, Malaysia.

37. Peter , J. P. (1981). Construct validity: A review of basic issues and marketing practices . *Journal of Marketing Research* , 18 , 133 – 145 .

38. Rajagopal , S. , Ramanan , S. , Visvanathan , R. , & Satapathy , S. (2011). Halal certification: Implication for marketers in UAE . *Journal of Islamic Marketing* , 2 (2), 138 – 153 .

39. Rossiter , J. R. , & Percy , L. (1998). *Advertising communication and promotion management ()* , 2nd ed. , New York , NY : McGraw-Hill .

40. Schiffman , L. G. , & Kanuk , L. L. (2004). *Consumer behavior ()* , 8th ed. . Upper Saddle River , NJ : Pearson Prentice Hall .

41. Shaari , J. A. N. , & Arifin , N. S. (2010). Dimension of halal purchase intention: A preliminary study . *International Review of Business Research Papers* , 6 (4), 444 – 456 .

42. Shah Alam , S. , & Sayuti , N. M. (2001). Applying the theory of planned behaviour (TPB) in halal food purchasing . *International Journal of Commerce and Management* , 21 (1), 8 – 20 .

43. Sparks , P. , & Shepherd , R. (1992). Self-identity and the theory of planned behaviour: Assessing the role of identification with "green consumerism" . *Social Psychology Quarterly* , 55 (4), 388 – 399 .

44. Sparks , P. , Shepherd , R. , Wieringa , N. , & Zimmermanns , N. (1995). Perceived behavioural control, unrealistic optimism and dietary change: An exploratory study. *Appetite* , 24, 243–255.

45. Sumali , A. (2006). Halal: New market opportunities. *Proceedings of the 9th Efficient Consumer Response (ECR) Conference*, 1–19.

46. Verbeke , W. , & Vackier , I. (2004). Profile and effects of consumer involvement in fresh meat . *Meat Science* , 67 (1), 159 – 168 .

Index

N

O

P

R

S

T